"Let us declare nature as legitimate. All plants should be declared legal. The notion of illegal plants in a civilised society is obnoxious, ridiculous and absurd."

Contents

The Merging of Nature and Technology

The term hydroponics was originally coined in the mid 20th century. It is a term used to express a technique for growing plants in a soilless medium.

The concept of growing plants without soil goes way back in time to our prehistoric past. For example, the mythical Hanging Gardens of Babylon, the much worshipped flooding of the Egyptian Nile and the Floating Gardens of Mexico City are all examples of hydroponics.

History, as it always seems to do, has turned full circle and the rebirth of hydroponics is back with us. Approximately 90% of all cut fresh flowers purchased in the UK are hydroponically grown and an estimated 65% of all fruit and vegetables purchased from your supermarket, again are grown in hydroponics systems. So like it or not, we have all bought and eaten hydroponic produce. In many countries, hydroponics is big business, but in the UK we were a bit slow to catch on. Ironically, a British professor invented, or should we say re-invented hydroponics in the Sixties. Then, hydroponics like a lot of very good British inventions was adopted by the rest of the world who developed and exploited it. However, the UK hydroponics industry has now established itself and is leading the way once again.

Plants are grown in an inert, sterile growing medium and fed a mixture of water and nutrient. The principle is basic. Plants that are grown in soil have to continuously develop their rootballs in search of water, nutrients and air so the majority of the plants available energy is spent on the lower root development restricting their upper growth. In hydroponics, water, nutrient and air are mainlined directly to the rootball, freeing the plant to use its available energy in its upper leaf, fruit or flower development. That coupled with the fact that it has all the specific nutrient, air and water it could ever want, means a plant can grow at a previously unheard of rate. If you like it's "super charged battery farming for plants on steroids" but the plants are in heaven, not hell. Because plants grown in a hydroponics system can be given very exact and specific doses of nutrient, a crop raised hydroponically will develop optimum levels of appearance, yield and flavour. Because the roots of plants grown in a soilless medium do not need to constantly grow in search of nutrient, more plants can be grown in a smaller area. You will thus be making the best use of whatever space is available to you.

Why Hydroponics?

Yield	2 to 10 times more than soil!
Nutrition	no deficiencies or toxicities
Water	$^{1}/_{10}$th that of irrigation
Diseases	no soil bacteria or viruses
Weeds	virtually none
Quality	higher, better, healthier
Costs	initially higher, then lower
Maturity	non-seasonal, faster!
Sanitisation	easy, safe, sterilisable
Transplanting	minimal problems
Environmental	compatible, safe
Convenient	grow your own!

In short, hydroponics allows you to grow approximately twice to ten times the yield, in half the space and in half the time. So, if you have a greenhouse, conservatory, spare room, loft, cellar, or closet, you too can become hydroponically enlightened.

There are few experiences more rewarding than cultivating crops; possibly mankind's oldest vocation. Experienced growers know what it is like to participate in the life cycle of plants, to create, to nurture, to encourage and to stimulate a crop to fulfil its potential gives you a special kind of joy and fulfilment.

A wise shaman once said - man's survival depends on three things; the merging of Technology and Nature, Science and Religion, and Dance and Idea. The trinity is now with us; let us embrace it.

The hydroponic experience will bring new meaning into your life, one that is mystical, enlightening, educational, satisfying and fun.

The Advantages of Hydroponics Over Dirt

No soil means that you can grow plants everywhere and anywhere you want. This would include big cities, towns, suburbs and the countryside. No matter if you have a garden or not, or that your soil quality is substandard, it allows you to grow where no one has previously grown before.

No soil means no soil-borne diseases, pests or weeds and as a consequence means no herbicides or pesticides.

No soil means huge savings in water and fertilisers as only the water and nutrients that are required get used by the plants when they need it, and so there is no wastage into the ground or via evaporation.

No soil means much less of the plant's available energy is spent in search of water and nourishment thereby freeing the plant to divert its available energy on growth and enlargement. For example, in soil the plant transforms its food into absorbable ions by composting organic matter. In hydroponics these same ions are already available directly to the plant so the plant has nothing to do except absorb and grow. Another example is that in soil, plants waste a lot of energy developing large root masses to find and absorb the nourishment they require. In hydroponics, the roots are submerged in a perfectly adapted oxygen enriched nutrient solution, so the plant does not have to spend much of its available energy on lower root development allowing the plant's energy to be focused on the upper foliage, flower and fruit development instead. This process results in a more viable use of space and visible improvement in growth and yields.

No soil means exceptional levels of gas exchange in the root zone, therefore, accelerating the development of your plant's growth and yields.

No soil means that you are allowed the opportunity to closely manage your plants' needs and therefore stimulate its growth. The optimal environment hydroponics offers promotes the very best utilisation of the plant's genetic potential. This process also allows significant shortening of the growth times and also production intervals.

No soil means the strength and vigour of plants started and grown on in hydroponics media is such that many soil growers will start their cuttings in hydroponics media during the winter, then transplant them outdoors in dirt during the good season gaining many weeks and even months growth on their outdoor crops, and allowing the crop to be harvested early when demand has not yet met supply.

No soil means that once you have set up your

hydroponics system, it will run almost indefinitely without much additional investment. Unlike having to always throw your dirt away and replace it, hydroponics is reusable and clean.

No soil, in short, allows you the opportunity when using hydroponics to gain a 30% increase in your crop, allows you to grow 30% more plants in the same area, allows you to grow the plants 30% faster, allows you to get 30% more active ingredients in your plants and most importantly, allows you to gain 30% higher yields.

Why?

So why has the world not become hydroponicised?

Is it because the layman thinks it is chemical i.e. artificial and man-made, therefore harmful to the consumer and the environment? In fact, the reverse of this statement is actually the truth, and in time the truth will be told so that the layman can understand, then the truth will have to be believed.

So what are hydroponic nutrients made of?

Are they hazardous chemicals or harmless and organic in nature? To answer this question properly, first it is important to understand the difference between organic and hydroponics and understand how plants feed. The difference between organic nutrients and hydroponic nutrients is based on the following: organic means that no unnatural or man-made chemicals are used to make the fertilisers, whereas in hydroponics the fertilisers are made out of purified mineral salts.

In soil, several millions per gram of earth micro-organisms decompose organic matter and transform organic molecules that are non-absorbable by plants into ions that plants can absorb.

In hydroponics as in soil, plants absorb their foods in the form of ions. So if you are to correctly define this, there is in fact no difference between an ion of organic origin than that of an ion from a mineral origin. They are in fact the same. These same minerals are absorbed in either hydroponic or organic growing.

In organic growing they are released from organic matter by the action of worms and bacteria via a composting process which in all truth is a chemical reaction.

In hydroponics, water soluble mineral salts in the form of the same ions provide these same elements.

So, with this in mind, is it artificial and man-made? Well it is, but only as much as the refined organic sugar that you get from the supermarket nowadays. So if you are willing to buy and consume this, then what is the problem?

This is a classic example of a little bit of isolated knowledge disguising the real truth and allowing an opening for the ill-informed to become righteous about something they actually know very little about. This in turn gives bad press to a subject that in all honesty is none deserving of it.

Hydroponics deserves to be evangelised, not demonised, which sadly seems to be the current stance of today's media.

The use of this technology allows you to grow where no one has grown before, be it within a starving nation's land whose soil can no longer sustain viable crops, be it underground or above, in space or under the oceans, this technology will allow humanity to live where humanity chooses. If employed for our survival or colonisation, hydroponics is and will be a major part of our collective future.

Chapter 1

Plants - A Basic Overview

Before you dive in at the deep end of this book, it is advisable that you gain a basic understanding of what a plant is and what a plant does. The most competent indoor gardeners can learn a lot from the next few pages so don't skip past thinking that you know about this bit, because most probably you don't.

Plants can be divided into 3 classes:

Cam – which covers most succulent plants; these tend to be low light-loving and high humidity plants.

C4 – which covers most grasses; these tend to be medium light-loving and good CO_2 using plants.

C3 – which covers most high energy plants; these are the flower and fruit bearers of the plant kingdom. High light-loving and excellent CO_2 using plants.

This book is based around the knowledge of C3 cultivation, which is indeed the most popular type of cultivated plant in hydroponics systems to date.

A C3 plant consists of two joined 3-carbon atoms to manufacture sugar. Sugar is 6-carbons, 6-hydrogens and 6-oxygens fused as one.

C3 plants manufacture energy through photosynthesis, which in simple terms is the way plants breathe and grow. The plant uses light, water, nutrient, and CO_2 in the atmosphere, then absorbs it and converts it into sugar, and in so doing, releases oxygen. The plant then moves around sugars, water and nutrients within its structure to create growth. The leaves suck up water from the roots through tubular cells called xylem. The leaves are constantly respiring, evaporating water supplied via the roots and this creates the water tension or pressure that keeps the plant rigid and strong. Leaves then send sugars via photosynthesis down to supply the roots through a tubular cell called the phloem. This creates a perpetual cycle with the roots supplying the leaves with the water they need so that the leaves can supply the roots with the sugars they require and a simple symbiosis is formed. The liquid, being in constant motion, maintains the strength and structure of the plant.

A C3 plant can be broken down into 3 main sections:

- the roots
- the stems
- the leaves

These three things are the battery, engine and fuel of the plant. If you develop a problem in any one these three areas, the plant will be in quite serious trouble.

The alchemy of this truth is "As Above So Below". Always remember this.

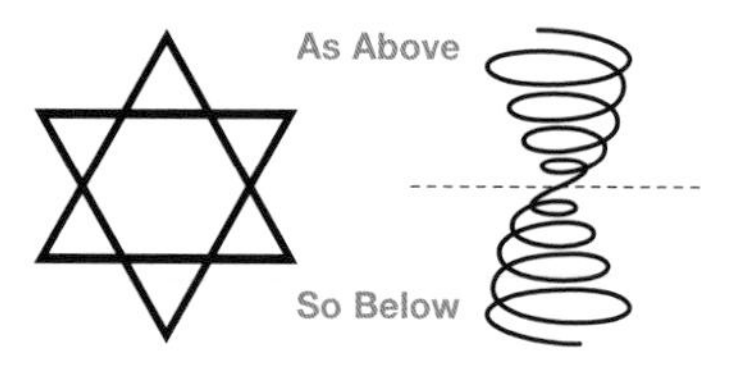

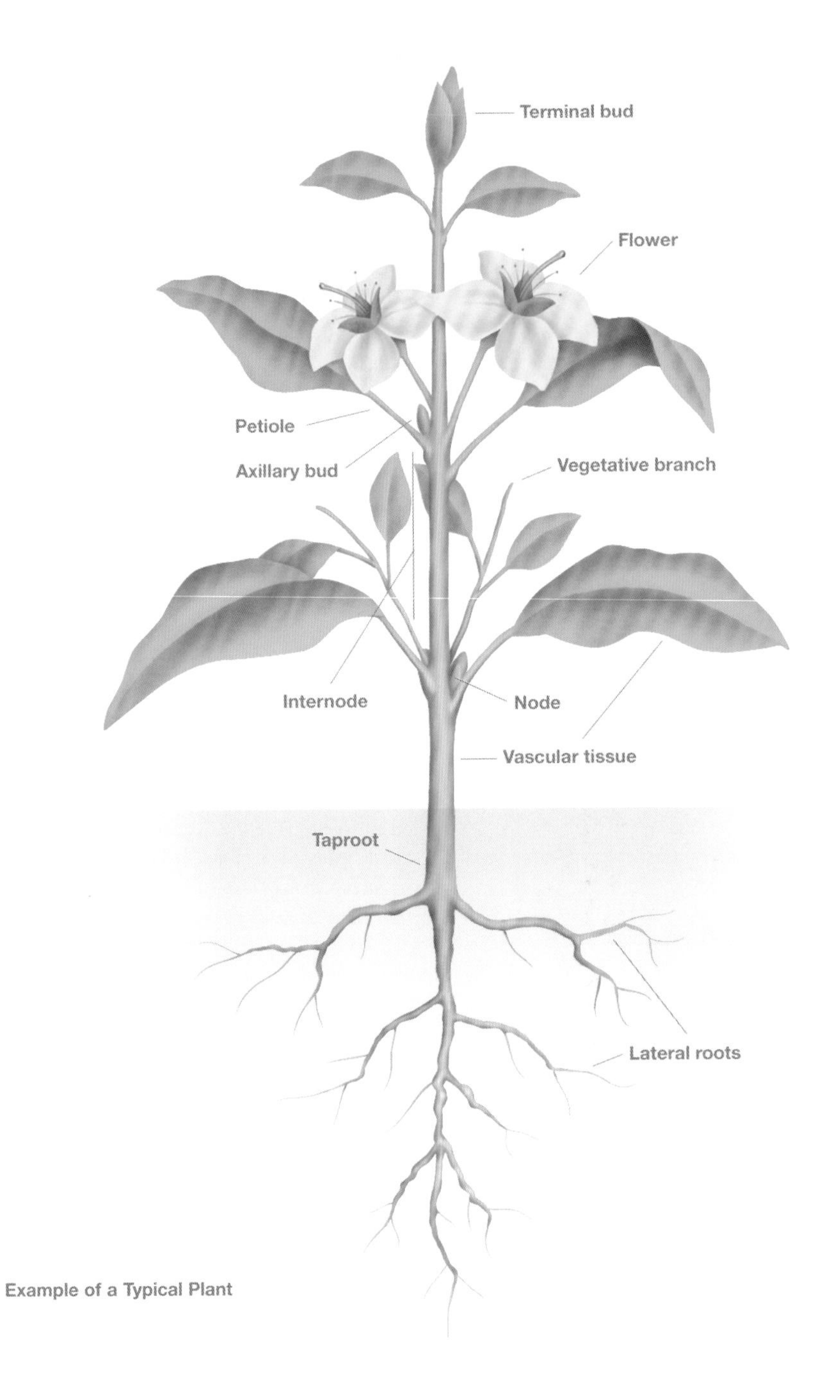

Example of a Typical Plant

Roots

Growth emerges first from the root; 'tis the first thing that grows. Roots pump nutrients and water to the leaves in exchange for sugar. The rootball is also a battery for the storage of excess energy created via the leaves, which gets stored as starch. This ability to store unused energy is the key to bumper yields. The healthier the roots are and the greater their capacity to store starch, the greater the plant will grow and the greater the yield will be.

Roots are organised in layers of cylinders. For roots in primary growth, the outermost cylinder consists of the dermal layer which is constructed with epidermal cells. You will also find root hairs which are the unicellular extensions in the epidermis which function in the uptake of water, nutrients and oxygen. The cortex is the next cylindrical layer. This is a layer of ground tissue which is made up of parenchyma cells. The endodermis cylindrical layer follows which is the innermost layer of the cortex. The endodermis is distinct from the rest of the cortex. The major distinguishing factor is the casparian strip, which is a layer of a waxy substance called suberin embedded into the endodermal cell wall. This casparian strip prevents the passage of water and molecules through the endodermis via an apoplastic pathway. Inside the endodermis is the pericycle which is a layer of meristematic cells which functions in the formation of lateral roots, the vascular cambium and the cork cambium. The last cylinder consists of a vascular which functions in the transport of water, sugars, and other nutrients up and down the plant.

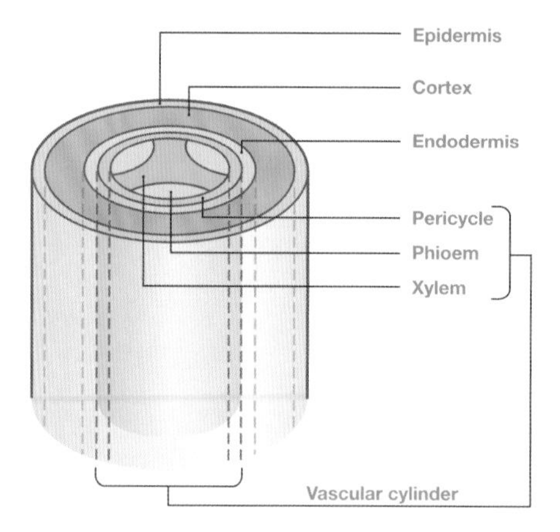

Example of a Cross Section of Root

The greater and more extensive the size of area that the roots have to grow in, the bigger the rootball can become, the better the plant will grow and the more the plant is able to yield. This is due to rootballs being able to store lots of energy and are therefore capable of exchanging many types of nutrients pumped up to the leaves, thereby allowing the leaves to pump down more sugars and allowing the cycle to progress exponentially. The growth of the rootball is directly influenced by water, oxygen, temperature, nutrients and the sugars pumped down via the leaves photosynthesising the light.

A classic example of this process is comparing a soil grown plant to one in a hydroponics system. In soil, the plant has to continuously develop its root system in search of water, nutrients and air, so with what available energy the plant has, it has to spend a substantial amount on its lower root growth, thereby allowing minimum amount of energy for its upper leaf development.

In hydroponics systems, water, nutrient and oxygen are directly mainlined to the rootball therefore freeing the plant from expending substantial energy in search of these. This in turn allows the plant to spend this energy on upper leaf growth, which in turn develops the roots.

Soil plants spend too much energy on lower root development prohibiting the plants' upper leaf development. In hydroponics, this energy is readily available for the upper leaf development, which in turn accelerates the root growth. In short, the less energy the roots use to absorb water, nutrients and oxygen, the more available energy they have to grow which creates greater and more extensive leaf systems, and in turn creates greater and more extensive rootballs.

Roots can again be divided into 3 sections:

Water Roots – Taproot - found at the bottom of a rootball.

Air Roots – Root Hairs - found at the top of the rootball.

Combination Roots – found in the middle of the rootball.

The top one third of the root system is comprised of specialised oxygen seeking roots. The bottom one third of the root system is comprised of specialised water seeking roots and the middle one third of the root system is comprised of both water and air seeking roots.

For this reason, it is imperative that the top one third of the rootball is not kept constantly wet as this will restrict aeration to the rootball creating a poor performing plant. Likewise, the bottom one third of the rootball should also never be allowed to dry out completely as this will again prohibit the performance of the plant. The middle one third of the rootball should be regularly soaked and then allowed to dry periodically.

Aeration is the key to maximum root development; the more oxygen the roots have, the better they will perform, the better the plant will grow and the better the plant will yield. Roots need oxygen to convert sugar to energy, which is then pumped back up to the plant, which in turn allows the roots to store more energy and so on and so forth.

It is not a widely known fact that plants do actually need to sleep. Most crazy US cultivation books rant and rave about 24 hour light cycles! What is this all about? and where in the world do plants flourish under a 24 hour light regime? The whole of the natural world needs a sleep so why would certain C3 plants be any different?

It has been scientifically proven that during night cycles, plants create vast amounts of their storage development. This is due to the fact that the roots are not being subjected to the demands that the leaves make to feed them more water, nutrients and sugars. Because the leaves are resting due to lack of light resulting in no photosynthesis, the roots can spend the available energy on storing excess energy for later use in the plants' life. Night cycles are critical for a heavy yielding indoor garden.

Stems

Stems are an integral part of the health and overall happiness of the plant. The stems act as a reflector indicating signs of stress and strength. Stems, due to their nature, are the supply pipelines for the roots and leaves.

The stem is the main supporting structure of a plant and serves to transport and to store water and food. The vascular system in the stem is mainly made up of xylem which are upward conducting, and phloem which are downward conducting tissues, usually in vascular bundles arranged concentrically on either side of the cambium with the xylem inside and the phloem on the outside.

The vascular bundle is a strand of conducting tissue extending lengthwise through the stems and roots of C3 plants. The vascular bundle consists of

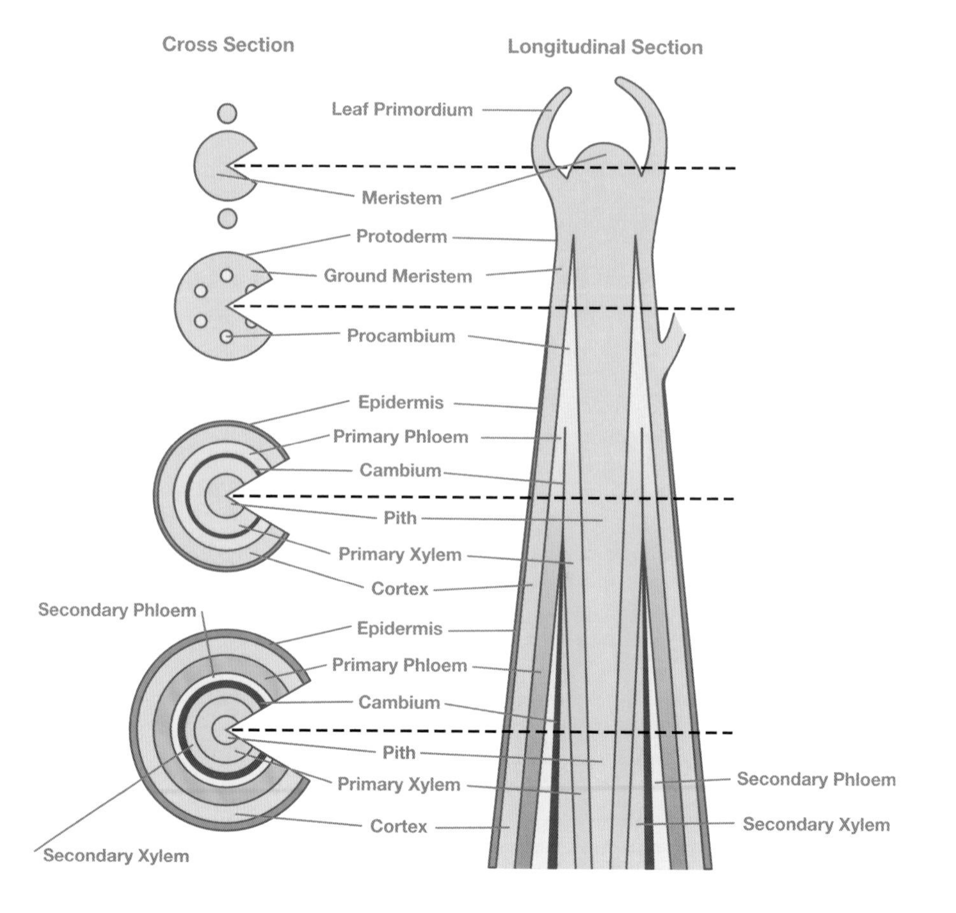

Example of a Cross Section of the Tip of a Stem

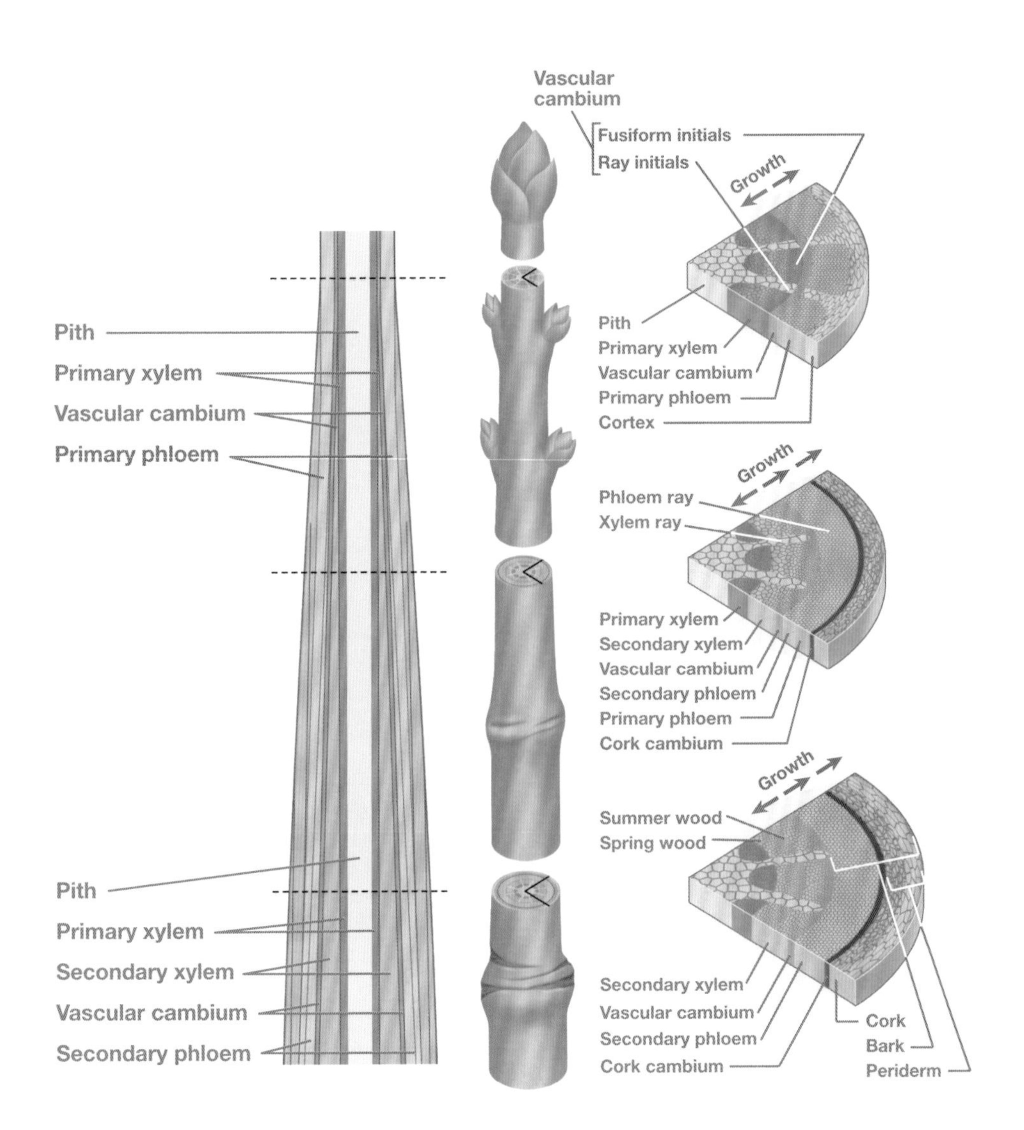

Example of a Cross Section of a Stem

xylem which conducts water and dissolved mineral substances from the roots to the leaves, and phloem which conducts dissolved foods, especially sugars, from the leaves to the storage tissues of the stem and rootball. The structure of vascular bundles varies among the different plant groups.

The pith is a central core of spongy tissue and is surrounded by strands or bundles of conducting xylem and phloem. The cambium, which is an area of actively dividing cells, lies just below the bark. Lateral buds and leaves grow out of the stem at intervals called nodes; the intervals on the stem between the nodes are called internodes.

The pith is the core of the stem of most plants. Pith is composed of large, loosely packed food storage cells. As the stem grows older, the pith usually dries out, and in some plants it disintegrates and the stem becomes hollow.

Plants contain two separate transport systems running side by side. These allow substances, such as water, minerals and sugars, to be transported to different parts of the plant. They also provide plants with support, helped by the distended cells.

Xylem is a continuous system of tubes running from the roots to the leaves. It consists of empty, dead cells with thickened sidewalls ... well this is the conventional view and for decades, researchers have seen the xylem as a column of dead tissue, like a worn pipe, that sits inside plant stems passively supplying water to thirsty leaves. However, in a recently published paper, a team of plant biologists reported that gels in key xylem membranes constantly shrink and swell. With this motion, the xylem actually adjusts the flow of mineral-rich water coursing towards leaves. The xylem is a part of the vascular system that transports water and dissolved minerals from the roots to the rest of the plant and may also provide mechanical support. Xylem consists of specialised water conducting tissues made up mostly of narrow, elongated and hollow cells.

The phloem is composed of various specialised cells called sieve tubes, companion cells, phloem fibres and phloem parenchyma cells. Sieve tubes are columns of sieve tube cells which have perforated areas in their walls and provide the main channels in which food elements travel. Phloem fibres are long, flexible cells that make up the soft fibres used commercially e.g. flax and hemp. The phloem may also be called bast tissues in plants that conduct foods made in the leaves to all other parts of the plant.

The cambium is a layer of actively dividing cells between xylem which are fluid conducting and phloem which are food conducting tissues and are responsible for the secondary growth of stems and roots, resulting in an increase in thickness.

The cortex is a tissue of unspecialised cells lying between the epidermis (surface cells) and the vascular or conducting tissues of stems and roots. Cortical cells may contain stored food or other substances such as resins, latex, essential oils, and tannins.

The epidermis is the outermost layer of cells covering the stem, root, leaf, flower, fruit, and seed parts of a plant. The epidermis and its waxy cuticle provide a protective barrier against injury, water loss, and infection. Various modified epidermal cells regulate transpiration, increase water absorption, and secrete substances.

The meristems are a region of cells capable of division and growth. Meristems are classified by location as apical or primary i.e. at the root and shoot tips, lateral or secondary i.e. in the vascular cambium and cork cambium, or intercalary i.e. at

internodes, the stem regions between the places at which leaves attach, and at leaf bases, especially in certain monocots, e.g., grasses. Apical meristems give rise to the primary plant body. Lateral meristems provide increase in stem girth. Injured and damaged tissues can convert other cells to new meristem for wound healing.

The stems react first to the beginnings of over fertilisation. The veins on the stems become a brightened reddish colour. This starts from the base of the stem then migrates to the upper levels of the plant's stem and finally through to the leaves, which in turn causes the tips to burn and the leaves to curl. Flushing with a lowered nutrient solution can easily rectify this.

Elongation of stems indicates one of two things and sometimes a combination of both; lack of light and heat build up.

Stems stretch as they look for more light and this indicates that the light source should be moved closer to the canopy of the plant or that the grow room is in need of more light sources.

Stems also stretch when temperatures in the grow room are too high. The reason this happens is that the plant creates more surface area to transpire through and cool itself down.

Sometimes, both these factors combined are responsible for stretched plants.

Stems can also reflect if young plants are being over watered as the stems at the base narrow then fall over. This is called damping off where the plant stems literally have rotted at the base.

Stems can also reflect that a plant is thriving; a very happy plant develops a stem which has a velvet glow encompassing the whole surface area of the stem and it also produces thick, bulging veins that harden into long ridges from the base to the tip of the stem. Over time this produces thicker and thicker stems creating a sound foundation for a very healthy plant.

Typically, the thicker the stem, the happier the plant is and the more weight and yield it is able to support and produce.

On that note, the shorter the internode is between the leaf joints, the more successful the plant will become as the energy supplied via the rootball up to the leaves has a shorter distance to travel. This allows the plant to conserve energy for the growing process instead of having to try and pump nutrient to greater heights if the stems are elongated, resulting in wasted energy.

Leaves

Leaves are the manufacturers of plant sugars. They utilise the light to combine water, nutrient and carbon dioxide to produce sugars which are then sent down the stem to feed the rootball. A weak link in any of these areas – light, water, nutrient, or CO_2 can prohibit the plant from producing as it should.

The undersides of the leaves have tiny little breathing holes called stomata. The stomata absorb the CO_2 in the atmosphere. They are the breathing apparatus of the plant. Approximately 30,000 of these holes can cover 1 cm^2 of the underside of the leaf. They are located on the underside to prevent dust, dirt, pollutants and spores from blocking them. In recent experiments, these stomata have been shown to swell and contract when subjecting them to differing levels of CO_2.

The leaves cause water and nutrient to be drawn up from the roots via the stem to the leaves. Excellent leaf development makes it possible to suck up lots of nutrients, water and air which in turn creates lots of excess sugar to be sent down to the root system. These leaves are constantly pushing out energy to the plant and the roots and when optimised, the leaves create so much starch that they begin to store the excess in the tissues of the leaf. During the night cycle this stored energy is diverted down to the roots. The healthier the leaves left on the plant, the better this process is, the healthier the leaf system is, the more energy it will create for further storage within the plant. The more stored energy, the greater the yield potential of the plant.

As with the stems, the leaves can be a good indicator as to the past, present and potential condition of the plant. The leaves reflect the health and happiness of the plant, however, leaves are unable to repair themselves, so damage done in the past is always visible. For example, if browning at the end tip of leaves and curling of the leaves outer edge can be seen, this is a good indicator that the plant has been over fertilised. The leaves tell you something is up so you back off on the amount of nutrient you are giving them until you are happy that you are no longer over feeding them. The new growth will look fine and indeed is the indicator of how well the plant currently is. The over fertilised leaves will lose the outer edge curl, however, they will always have signs of tip burn.

This is also the case with pest problems and disease. Even if the problem has been resolved, the aftermath of the situation is forever left as evidence on the plant. However, as new leaf systems grow, the old ones end up lower down the plant, and the new become the dominant system. So to recap, if new leaves and growth are without problems, the general health of the plant is well.

The leaves at the bottom of the plant are mainly used to feed the root system and the leaves at the top of the plant are mainly used to feed the new shoots, fruit or flowering sites. The leaves in the middle of the plant feed the plant's needs either way depending on the condition of the plant and what cycle the plant is in.

The more leaves that are able to collect light, the bigger the root system will become and the more stored energy will be left in the plant. This store of energy gives the plant the best chance of producing greater yields, therefore, do not strip any leaves from the plant. In recent years, techniques have been published such as the 24 hour light cycle and the stripping of leaves, especially the larger ones, which apparently then allowed for greater fruit or flower formations. This, like the 24 hour light regime is a total myth and should be disregarded as utter nonsense!

However, leaves are subject to entropy like any natural thing, so when the leaves have matured and turn yellow and are ready to die then these yellowing leaves may be removed. It is good practice to remove the older dying leaves in one fell swoop. Every time you subject a plant to a cut, it goes into shock for a few days and the plant will not do any growing during this period. So if you cut one leaf a day over the course of a week, it will not grow at all over this period of time.

Illustrated overleaf: the cuticle is a waxy layer that covers the epidermis of some leaves which reduces water loss. The upper and lower epidermis is the surface layer of a leaf that protects the inner parts of the leaf. The palisade layer is a layer of spongy cells loosely arranged between the palisade layer and the lower epidermis of the leaf. The chloroplast dislikes structures in cells that contain chlorophyll. The nucleus is the part of the cell that contains material that controls and activates the cell. The vacuole holds water. The air space holds air and CO_2 between cells.

Never remove dark healthy leaves as these leaves are in full swing production for the plant's needs. Removing these leaves will curb the available potential of the plant resulting in under yielding crops.

The majority of leaf development is created during what is known as the vegetative cycle of the plant's growth. This cycle is normally related to the photoperiod of 18-20 hours lights on and 6-4 hours lights off. It is possible to sustain a plant predominantly in a vegetative state by subjecting the plant to this photoperiod. A mother plant for example would be kept perpetually on this cycle in order to never allow it to fruit or flower so that clones can be made at any time the gardener wishes. When you reduce the photoperiod down to 12 hours lights on and 12 hours lights off, the plant is tricked into flowering, however, for at least 2 weeks after the switch to 12/12, the plants would still be in vegetative growth.

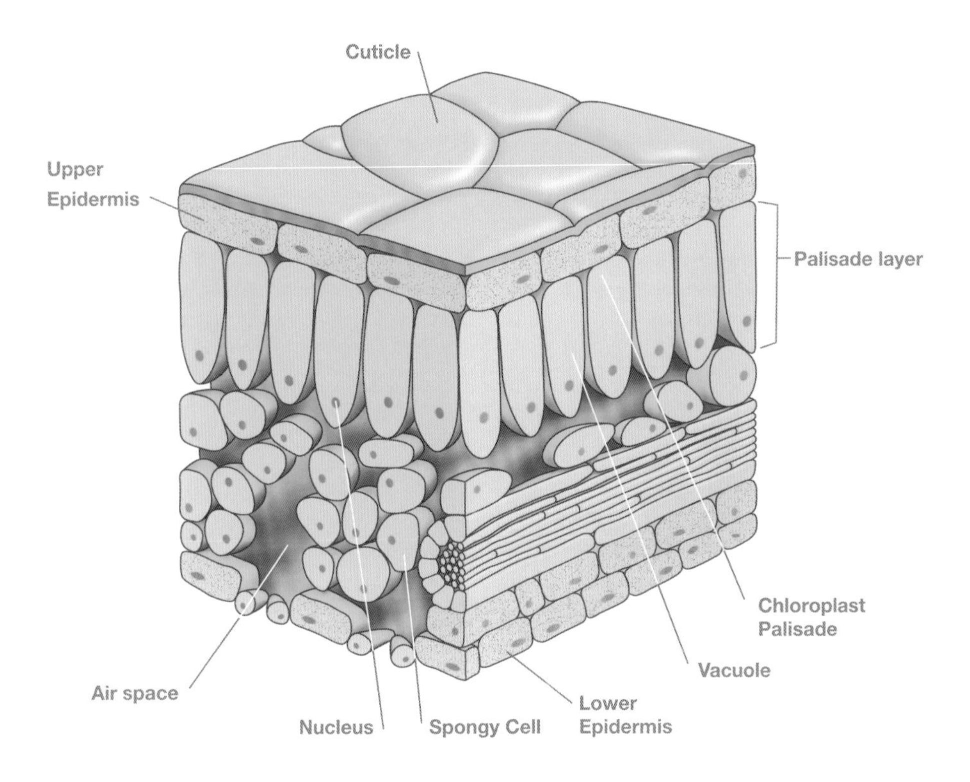

Example of a Cross Section of a Leaf

Flowering

A plant produces flowers for one primary reason and that is to reproduce. As the plant grows over time, and if the conditions are correct, the plant produces buds which in turn, becomes flowers. The flowers fade and fruit or seeds develop. This is nature's most fascinating and vital cycle with fertilisation resulting in reproduction.

In the plant kingdom, there are three basic types of flowers. Male flowers have only male parts and produce pollen. Female flowers have only female parts and produce fruit. Perfect or complete flowers have both male and female parts and produce both pollen and fruit. Please note that perfect or complete is their technical name in breeds of plants that naturally do this like cacti for example. However, in breeds of plants where this is not naturally occurring, for example in hemp, then these flowers are called hermaphrodites. In order for the seed or fruit to develop, the pollen produced by the male flower must be transferred to the female flower. This transfer is called pollination.

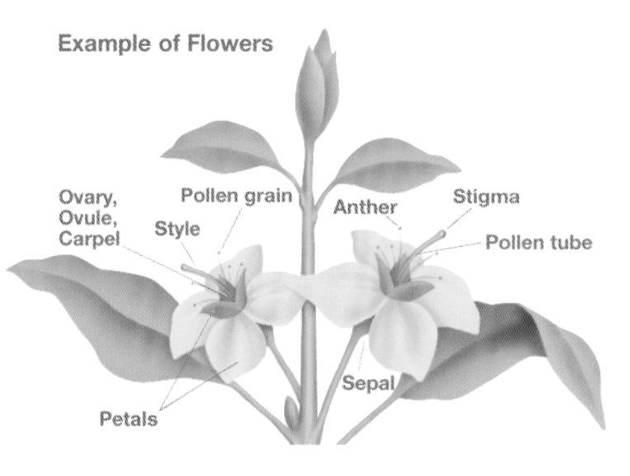

The diagram shows complete flowers to clarify this process. The sepal is one of the scales that makes up the calyx which is the outside casing of the bud. Within the calyx are the petals, composed of diverse forms and colours and having a dual function: firstly to safeguard the inner parts of the flower from injury until pollination can take place and secondly, to lure bees or other insects to the inner parts of the flower. Next are the stamens or the male parts of the flower and on the apex of each stamen is a pod-like anther which produces the pollen. When the pollen is ripe, the anther unfurls and the pollen is released. The pollen is transferred from the anther to the stigma which is the sticky part of the pistil, or inner section of the flower.

When pollen makes contact with the stigma, it germinates, extending down the ovary where it fuses with an ovule and then develops into a seed.

If the pollen transfers from the anther to the stigma on the same flower or the flowers of the same plant, and even plants of identical genetic material, then this is called self-fertilisation.

As mentioned earlier, flowering occurs when the plant is ready to try and reproduce; this is normally governed by lighting cycles.

A plant is more likely to flower as the daylight hours draw in, signalling that the summer time is drawing to an end. By reducing the photoperiod to 12 hours on and 12 hours off, the plant believes summer is coming to an end therefore wants to reproduce in order to uphold the survival of its genes.

The chemical mind of a plant is a lot slower to realise environmental changes than that of the mind that we possess. Always remember this and try to look at everything you do through the plant's perspective and not yours. A good example is the flowering cycle – it takes the plant at least 2 weeks to realise the lighting cycle has changed before it starts to flower. It is only due to this repetition of the light at the same time, night and day, over the course of 2 weeks that it now knows to flower; it recognises the environmental conditions have changed however, it only knows this through synchronised repetition over a 14 day period.

Plants do have a chemical mind per se, but this mind's consciousness is on a level that is hard for us to comprehend.

Similar to a dragonfly that lives 12 hours in our eyes, yet in the dragonfly's eyes those same 12 hours actually represent 70-80 years in the dragonfly's own perception and consciousness. The same applies to plants. If it is a fruit or flower bearing plant with a vegetative and flowering cycle which in total adds up to 7-9 weeks, each week if you will, represents approximately 10 years of life experienced by the plant. So each week in our eyes is actually approximately 10 years in the perception of the plant's reality. When you get your head round that perspective you bring yourself closer to the wants and needs of your favourite plants.

The flowering process of the plant normally takes approximately 6-9 weeks depending on the species. The start of the 12 hour cycle initiates flowering, so 6-9 weeks from switching to the 12 hour cycle, the plant should have matured and the harvest is ready for the picking.

Sometimes, the plants can be slower to mature and this often results in bumper crops, so do not worry if this is the case with your plants.

During the flowering cycle the plant will develop more leaf systems to sustain the fruit or flowers. Sometimes sucker systems can develop; these are shoots springing from the root or underground stem. These suckers are normally long, spindly and elongated. Remove these from the plant as they can tap a lot of the usable energy from the plant and can restrict large yields, as much energy is needed to maintain them.

Do not crop the plants early as this can result in a definitive lack of yield. It can also create less flavour and potency of aroma. A lot of the time, plants can develop second and even third flushes of fruit and flowers resulting in a plant producing a lot more than was ever realised. So do not crop them early; if in doubt let them be. Your fruit or flowers will then fill out, fatten and grow larger, normally resulting in a stronger, bigger, more flavoursome end product.

Hydroponics
Indoor Horticulture

Chapter 2

Plants and Entropy

Seedlings

Ideally, plants that are to be grown in a hydroponics system must first be rooted in a sterile substrate like rockwool. This would be the puritanical hydroponicist way of doing things and the correct way, however, you can bend the rules as we in our experience have discovered. Although not recommended, seedlings can also be raised in the traditional method and transferred to a hydroponics system. If you are to pursue this method, it is a good idea to wash the majority of dirt from the rootball and to dip the remaining roots and dirt in a weak hydrogen peroxide solution. Due to the differing strengths of H_2O_2, it is advised to refer to the manufacturers' instructions for the dilution ratio.

The benefits of using a sterile substrate is just that! It is sterile, therefore, you will not introduce any bacteria, disease or pests into your hydroponics system. This objective is sought after when building and maintaining a hydroponics system. The less variables to influence the system, the less chance of failure, disease or pests.

There also presents itself another favoured technique for raising seedlings and that is the use of Jiffy-7s. These are compressed peat pellets that when submerged in water expand to around 4 times their original size. These are preferred due to their ease of use; just add water and you are ready to go. However, although many do use these little wonders in hydroponics systems, they can contain disease, bacteria and even pests.

Using good reliable seeds normally results in an approximate 90% germination ratio. If you are going to store your seeds, then this is best done in a light-tight and waterproof container and stored in a cool or even cold area. If you have the space, a fridge is a great place to keep your seeds at optimum freshness. Out of the approximate 90% of the seeds that have sprouted, for the most part, as you are now dealing directly with nature, you will normally get a mixture of super strong seeds, strong seeds, average seeds, weak seeds and not forgetting the runts of the pack.

This is nature's way and to be honest it obviously works or we would not be here talking about it. This is the one drawback from raising seeds is that you will get inconsistencies and varying growth rates. However, all hydroponicists should know and experience how to germinate and raise seeds. Stock plants and mothers are not normally shared, so to be able to raise your own is fundamental, and this starts with the germination of seeds.

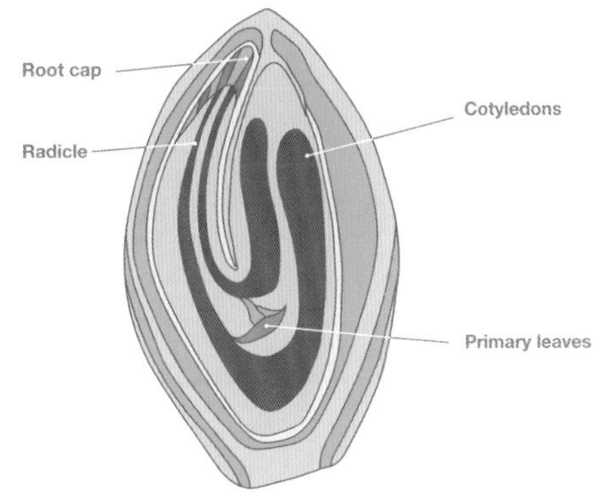

Example of a Cross Section of a Seed

The anatomy of a seed can be likened to an egg or even an embryo. The seed has built into it the genetic programming and genes of its parents.

Inside the seed, nearing the centre, are the primary leaves. Working out from the centre you have the cotyledons, then the radicle and nearing the top, the root cap. These are the four elements within the seed that create life.

Original genus seeds are seeds obtained from parents that have never been crossed. These are also known as pure genetic strains, normally imported in from countries that have been growing the same species of plant for hundreds, or thousands of years, and longer.

F_1 seeds are crossed varieties that have been stabilised. The stabilisation occurs due to a breeding program subjected on the parents over many generations. These are the most favoured seeds as the F_1s have the benefits of each of the new strains bred into the seed, i.e. vigour, flavour, growth rate, so on and so forth.

F_2, F_3 and F_4 are less stable. F_2s are obtained by fertilising F_1s and the offspring become F_2s resulting in less stable seeds, reverting back to the differing parentage that went into the making of the F_1s. The same applies when reproducing from F_2 or F_3 seeds in that the end result will be that the original seed stock that the parents came from was a complete mix which many F_1s are, and most probably would result in getting ten completely differing varieties from ten identical seeds. Sometimes this can be a blessing and sometimes not.

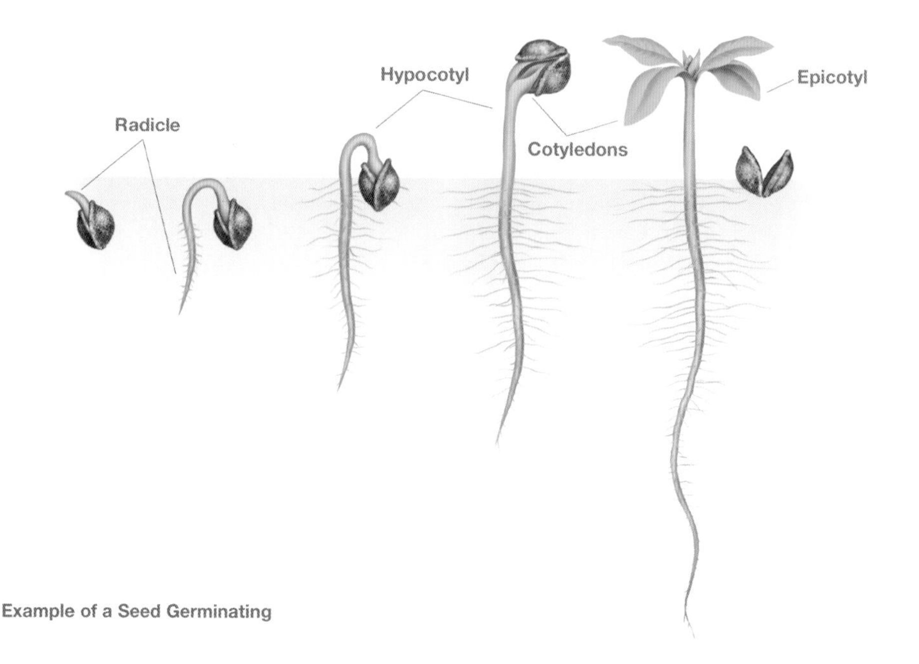

Example of a Seed Germinating

Cuttings

Cuttings, also known as clones, have been taken from plants since the beginning of time. Some species of plants are regarded as cultigens which means that they rely solely on mankind's intervention to take the cutting in order to reproduce. You can see that these plants over millennia have been cloned to the extent that they cannot reproduce through the normal methods, because generation after generation of clones have been taken from this particular species of plant and the genetics have adapted and know to rely on man to reproduce their species. This is a very strange thought indeed and a very interesting quirk in nature.

This proves that the relationship between mankind and plants is indeed symbiotic. Nature and man is a two-way exchange or at least it should be. ☺

We nurture nature and nature nurtures us. This is the old way and all indigenous cultures adhere to this belief. This belief should also be adopted and related to your hydroponics garden. You, the plants that you are growing, and the technology you are using is a symbiotic relationship. Together we, them, and it evolves.

The benefit of cuttings over seeds is that the cuttings taken from a F_1 mother for example will be nothing more or less than exact gene replicants of the mother. However, this can vary if you were to take ten cuttings from the same mother and take those ten cuttings to ten separate locations. Due to slightly varying environments, you can indeed end up with ten different results. The cuttings will adapt to the individual environments, which in turn will bring out the distinct traits from the differing parentage that went into the mother.

If the cuttings were taken from the same plant and grown in the same environment then the results will be pretty similar and sometimes identical. Again, I must remind you that we are dealing with nature here so that statement can be bent a little. An example is if you take 20 clones from a mother, some clones are taken from the top of the plant which have always been bathed in plenty of light resulting in that particular part of the plant having it pretty easy all its life. Then some clones were taken from the bottom of the plant beneath the canopy where the shoots have been struggling to get to the light and have had it pretty hard in comparison to the top of the plant. These two separate sets of cuttings will again produce slightly differing plants, due to the genetic make-up of one shoot having it easy and one shoot having it hard.

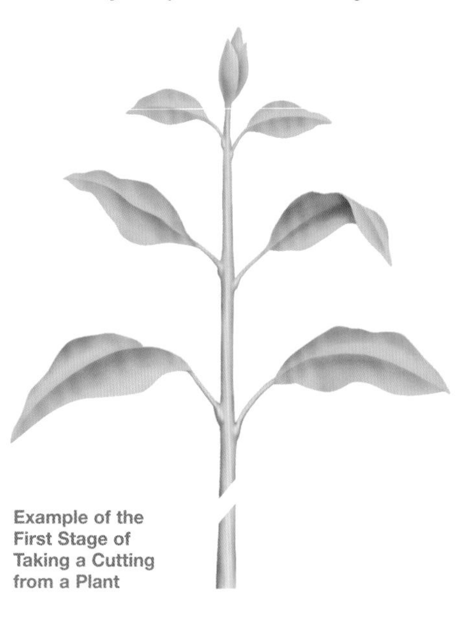

Example of the First Stage of Taking a Cutting from a Plant

Strangely, the shoots from the lower canopy make better cuttings than their counterparts from the top of the canopy. This might be because they have had to struggle to exist and fought for their right to grow making them stronger. Then, when given the opportunity to become a clone and go it alone, they prosper as they have more genetic will to survive, due to their underprivileged upbringing. So cuttings from the bottom canopy have more inertia to compete and survive compared to that of their counterparts from the top, where in comparison, these shoots have had it very easy all their lives, so when they become clones, they too are of similar genetic conditioning; that is in this case, they do not have the same will and inertia to survive and prosper, compared to that of their counterparts from the lower canopy.

Young Plants

To define a young plant is to refer to a plant that is beyond the seedling/cutting stage but is before the fruit or flowering stage. In a hydroponics system a young plant is predominantly in the vegetative cycle of the plant's growth.

Young plants like seedlings and cuttings are delicate and fragile. The seedling and cutting stage is the most troublesome with regards to stressing or damaging the would-be plant. The young plant stage is much more robust, however, it is still not robust enough to endure too much bad management.

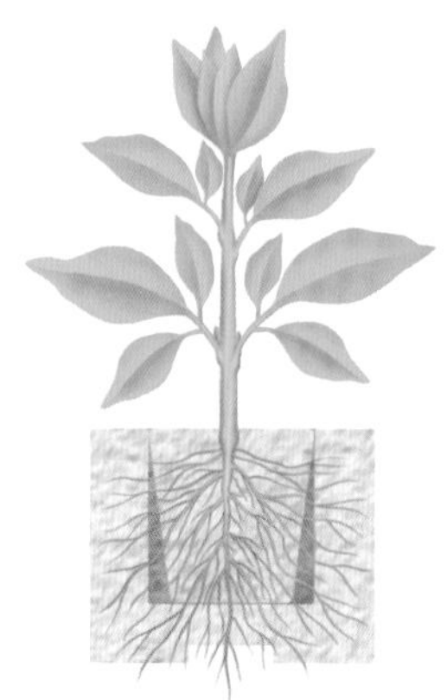

Example of a Young Plant

Some growers have been known to employ further propagation in the hydroponics system itself. This is done to minimise shock during the transition of transplanting from the propagator to a hydroponics system. The end result is that the plant takes quicker to the new environment and incurs much less stress than compared to being planted straight out of a propagator and into a hydroponics system. Although the young plant is really now in no need of further propagation, it has been proven that the young plant will develop its roots and adapt to the new system a lot quicker than without propagation, and furthermore can endure higher levels of stress.

So if you get it wrong but you are propagating within the hydroponics system, the plants will not react as violently as if it would without the further propagation. This technique ensures that you do not kill or damage your young plants and allows you an adequate margin of error. Further propagation of young plants in hydroponics systems normally only last for 7-10 days.

Once a young plant has taken hold in your hydroponics system and is showing rapid growth rate, it won't be long until it has become established within the system. After the plant has established itself and the growth rate has become exponential, then you are almost through the young plant stage.

The transplantation of seedlings or cuttings to a hydroponics system is again a time where it is possible to come unstuck and destroy or badly stress your beloved plants. It is crucial that you get all variables right before transplanting your seedlings or cuttings into your hydroponics system. Make sure that the day and night time temperatures are correct, that the CF and pH levels are right, that the watering regime is correct and that the lighting levels are also right. Get any one of these variables wrong and it could be kick the bucket time for your potential plants.

It is obviously wise to test run your grow room for at least one week or even two so that you are able to stabilise and make any adjustments to the environment and systems while no plants are in the location. This will also have the effect of buffering the hydroponics system. On almost all new hydroponics systems, it takes days and sometimes weeks to stabilise the CF and pH ratio within the system. It is therefore advisable that you run your system without your plants in it for a period of time in order to allow the system to buffer, so the plants do not suffer the shock of pH and CF fluctuation.

Establishing a Plant

It is fundamental that a plant is allowed to establish itself before you initiate forced fruit or flowering. If a plant is too young and has not been properly established in a hydroponics system or environment, you run a very high risk of either stressing the plant, resulting in a sickly and weak specimen, or that you end up growing a bonsai plant. Ironically, some growers wish this on their plants and are actually looking to bonsai them. This is not a very productive technique as you need to grow these in great numbers to gain good yields.

Advanced hydroponicists are from the school of thought that less is best and that less yields more.

One perfectly grown and large hydroponic plant, can yield that of 20 badly grown ones.

This whole book is dedicated to the process of maximising the yield available to each individual plant. Therefore to recap, it is very important to allow the young plants to establish themselves in the system and the environment, before initiating fruit or flowering cycles.

Some growers tend to overdo the vegetative cycle to allow the development of a good strong rootball. Now, because the plant has grown in height and size before they flip over to the fruit or flowering cycle, they effectively cut the plant size down in half, then allow a couple of days for the plant to get over the shock, then initiate the fruit or flowering cycle.

The benefit of this is that the plant has developed a big rootball which works as a supercharged battery to fire the plant's growth. The plant is well and truly established but is now only the size of a young plant, and the result of halving the height of the plant results in multiplying the amount of possible fruit or flowering sites. If one is to adopt this technique, it is critical that the plant is allowed to get over the stress of the cutting back, but also that the plant is never reduced by more than half of its mass. If you cut back any harder than half, you again run the risk of excessive shock and the possible production of bonsai plants. Most hydroponicists will cut the height back only and not the foliage, i.e. everything at a pre-defined height level is cut back but nothing past that particular height.

Please note that this technique is not stripping, it is solely the reduction of the plant's height.

Wise growers rarely practice the technique of stripping, that is unless subject to grow room constraints and this particular author frowns upon it.

The plant needs the leaf as much as it needs the roots and stem. However, the cutting back of established plants allows the plant access to the great reserve of energy held in the rootball, so that when the plant calls upon this reserve during the fruit or flowering process, the results are explosive blooms and a much heavier yielding plant.

Maturing a Plant

This particular area of study is very important.

When harvesting a plant before its time this can result in a definitive loss of yield, flavour, potency and aroma.

When harvesting a plant too late after its time this too can affect its taste, potency, flavour and aroma. When you harvest at just the right time, the best end product is realised. If in doubt, it is best advised to crop your plants late rather than early.

It is advisable to harvest the plant's fruits or flowers just after two thirds of the plant's fruits or flowers have changed to the ripened colour of that particular fruit or flower. So in effect, approximately one third of the fruits or flowers are still to ripen, but the majority of the plant's fruits or flowers are indeed ripe. At this point in time the plant is at its most flavoursome with maximum potency, aroma and taste.

If you crop your plants before this time, the plants can have a lot less flavour, potency, aroma, taste and weight. Also if you crop your plants after this time, the plants can also start to lose its flavour, potency, aroma and taste.

The later you crop your plants the more you increase the likelihood of bacteria or fungal problems like mould. Only in cases of these types of problems should the plants be harvested immaturely.

In an indoor garden, some species of plants are known to be able to reproduce a second and third time and sometimes more. This is known as regeneration. This is rare but not unheard of. The grower removes the mature fruit or flowers then re-initiates the vegetative cycle by increasing the photoperiod to 18-20 hours on and 4-6 hours off. The grower then grows the plant on this cycle for 2 to 3 weeks so the plant re-establishes itself. The grower then initiates the fruit or flowering cycle again by reducing the hours of light down to 12-12; the plant then flowers again.

If the plant is up for it, this can be done again and again.

However, most species of plant only have it in them to fruit or flower the once before they give up as all their energies are put into the fruit or flowering process.

If the plant looks green and healthy after you have cropped the fruit or flowers then indeed you should be able to regenerate the plant. This is rare although some plants can have it within their genetic make-up to do this. When this does occur, the second crop in terms of yield can surpass the first. However, although debatable, the flavour, potency, aroma and taste are argued to be not as desirable as the first crop.

So to recap, do not crop your plants too early or even too late; the plants will tell you when the crop is ready and if it is possible to regenerate them or not. If it is possible to regenerate, then it is a course of action worthy of pursuit.

Chapter 3

Propagation

Seeds

There are a number of options available when germinating seeds. We will focus on each technique one at a time.

Pre-soaking

The traditional pre-soaking method is a very simple way of getting your seeds to sprout. Simply get two sheets of very damp or almost wet kitchen roll paper towels. Wet them using normal tap water or bottled mineral water. Place your seeds between these two sheets of kitchen paper towels, then place these towels between two saucers or small plates, making sure you get a tight join so you do not lose any humidity inside the area between the two saucers or plates. Check daily, never letting the kitchen roll paper towels dry out. After 1-4 days, the seeds would have sprouted a taproot and this pre-sprouted seed can then be planted into a growing medium.

This method is good for those stubborn seeds that will not pop using other methods, which we will expand on shortly.

The only drawback with this method is the seedlings can suffer massive transplant shock and even possible damage to the taproot when removing from the kitchen roll paper towel, and then planting into the new medium. It is very easy to damage the seedling when actually picking up the seed with fingers or tweezers. However, when a seed won't pop using the other methods mentioned below, then this is your best and last resort. Even the most competent growers have found, using rockwool as an example, that for all their experience and previous success, that somehow this time, the seeds won't pop in the rockwool substrate. So these growers then resorted to the method mentioned above and within 24 - 48 hours all the seeds had finally popped ready for transplanting back into the rockwool for further propagation.

Rockwool

Modern rockwool methods are regarded as the most professional way of germinating seeds. This method is becoming more and more the norm used by nearly all hydroponicists. It involves using pre-soaked rockwool substrate; this is not normal rockwool but a specific one developed directly for horticulture. The most popular brand of horticultural rockwool is Grodan.

The process is simple but ideally requires using specialised hydroponic electronic equipment. A digital CF meter and a digital pH meter is advised, available from all good grow shops. Briefly, the CF meter measures the conductivity factor of dissolved salts in the water. In English, this gadget measures how strong or how weak your nutrient mix is, giving you a yardstick to understand if you are possibly over or under feeding the plants. The pH meter is self-explanatory; it simply measures the current pH level of your water or nutrient mix. This again is invaluable, as this will give you another yardstick in understanding and maintaining the right pH level for your plants.

Example of how a CF Meter Works

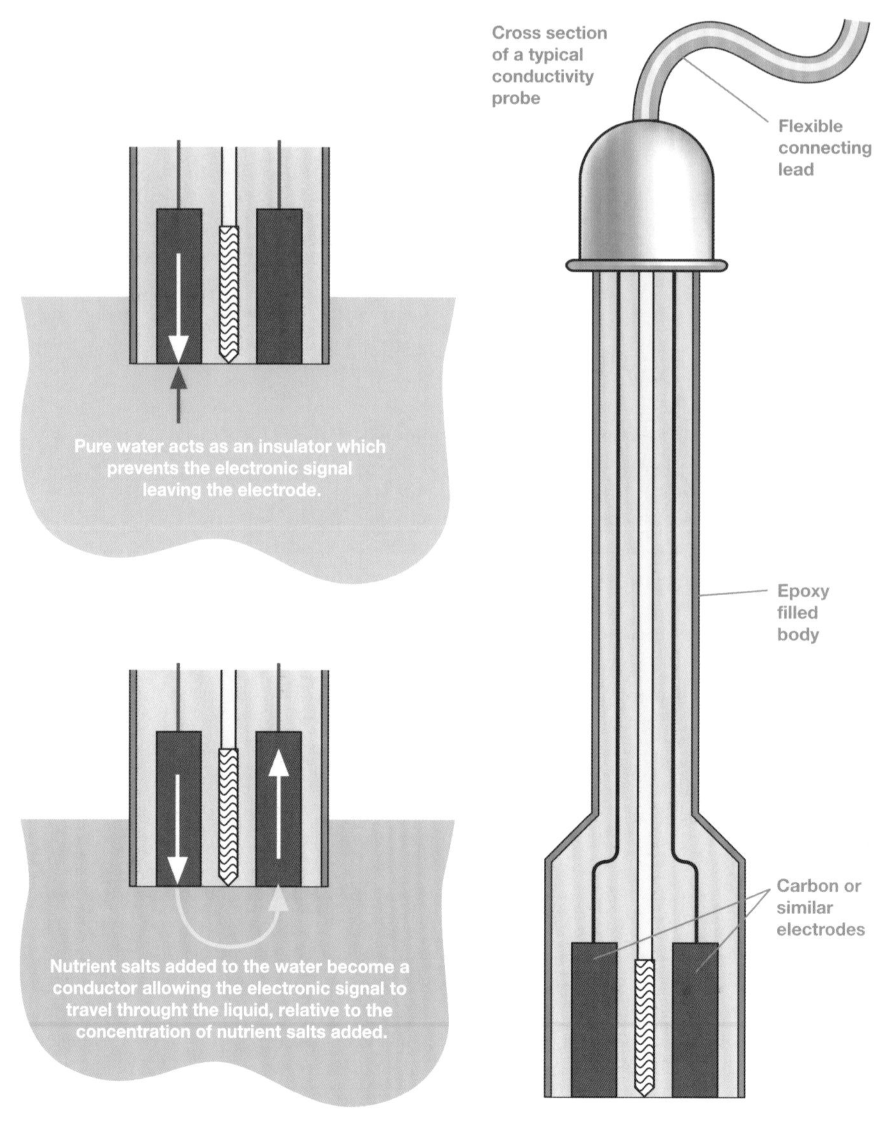

Firstly, you have to make a stock solution for your rockwool to be pre-soaked in before you plant your seeds. Using normal tap water, simply fill up a bucket or container. The next step would be to add some weak nutrient so the rockwool has food in it, so when the seeds start to grow they can get nourished quite easily.

If using normal 2 pack hydroponics nutrients, it is advised to start with only making this nutrient solution one quarter strength from the directions given to you on the side of the bottle. As an example, Canna nutrients are possibly the best selling nutrients available. These nutrients are stated to be diluted at 20 ml of A and 20 ml of B per 5 litres of water. So if you were to mix this nutrient for seeds, you are advised to knock it up at one quarter that strength which would be 5 ml of A and 5 ml of B per 5 litres of water.

At this point the CF meter comes into its own as this little beauty can now verify, if or if not, you have the desired CF value for your seeds and seedlings. A low CF value is required when germinating and raising seeds; it is recommended that a CF value of between 8-12 is reached.

If you find that the CF level is too high, then just add water to dilute it down until you get to the right level. If you find that the CF level is too low, then add equal amounts of A and B solutions until you have reached the right level.

Please note, nutrient solutions take time to mix and dissolve into the water, so don't be hasty, the nutrient might just be taking its time to mix and dissolve. Once you have established that you have the right CF level then it's time to balance and buffer the pH of the solution.

It is always advisable that the nutrient level is balanced first, as the nutrients can affect the pH level of the water you are adding it to. If you balance the pH first, you might have to balance it again after adding the nutrient. This can be time consuming and is also not recommended for the plants, as the more pH up or down that you use, the worse off the plants will be as both pH up and pH down attack vital food chains that are in the nutrient solution, thereby making it unavailable to the plant when it needs it. pH down is normally phosphoric acid and pH up is normally potassium hydroxide. In recent years a new pH down has come on to the market which is a blend of nitric based acids. It is claimed that this product is a food friendly based acid which will not damage any food chains like iron, zinc or calcium. However, if normal pH up and down are used correctly, then the nutrients or the plants will not mind it at all, they will actually greatly benefit from the process of having the pH maintained at the correct levels.

One word of warning; never, ever mix pH up and pH down together directly as a concentrate as these two chemicals are highly aggressive and react violently when mixed together.

Once you have mixed the nutrient and are happy that you have it at the right level, then it's time to balance or buffer the pH. The rockwool likes to be

Nutrient Solutions Comparative Table

C.F.	2	4	6	8	10	12	14	16	18	20	22	24	26	28
E.C.	0.2	0.4	0.6	0.8	1.0	1.2	1.4	1.6	1.8	2.0	2.2	2.4	2.6	2.8
T.D.S.	140	280	420	560	700	840	980	1120	1260	1400	1540	1680	1820	1960

10 x C.F. units (C.F.10) = 1 x E.C. unit (E.C. 1.0) = 700 parts per million (700 P.P.M.)

pre-soaked in a nutrient solution with a pH level of or around 5.5. This is the ideal pH level for rockwool to be pre-soaked in. As the rockwool is slightly alkaline, you need to make the pH solution more acidic to counteract the consistency of the rockwool. It is ideal if the seeds are germinated in a neutral pH level, and when reducing your pH level to 5.5 you are actually creating a neutral environment within the rockwool with the low pH counteracting the naturally high pH of the rockwool medium.

water area and you are adding pH down to your solution, if you add too much pH down into your solution then you are best advised to throw it away and start again rather than trying to get the level back up with pH up. This is due to the fact that pH up and down are aggressive and when administered to the solution they will attack your nutrient. Adding up to counteract the down is obviously bad practice as you have administered too much in the first place, then you are administering more which does your nutrients, and for that matter, your plants no favours. Same applies if you are in a soft water area and you are administering pH up to your solution and again you administer too much pH up, then you are again advised to throw that solution away rather than to try and balance it out with pH down. Less is always best in the world of hydroponics.

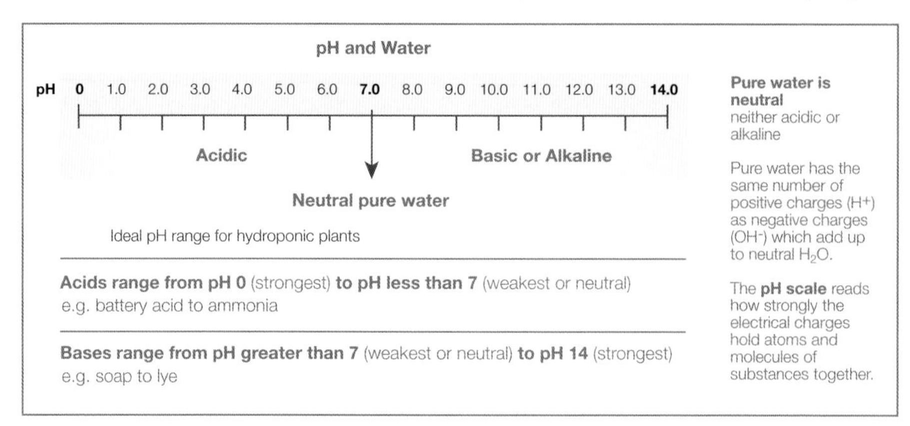

After adding your nutrient to the solution and adjusting it to the right level, it's time to test for your pH level. If your meter reads high you need to add pH down to it, and if too low then you will need pH up. Both pH down and pH up are extremely strong and very aggressive, so handle with care and only administer a tiny amount at a time to your solution, mix your solution well and allow the pH up or down to dissolve, then take another reading using your pH meter. If you find the value is still too high or too low then re-administer another tiny amount of your pH up or down.

If you find that you administer too much pH up or down then it is best practice to throw your solution away and start again, i.e. if you are living in a hard

Once your solution has been buffered with the right level of nutrient and balanced at the right pH level, it is now time to soak your rockwool blocks. Some growers allow the blocks to soak overnight, and in practice is the right procedure when using tap water as this allow the chlorine to rise to the top of the water, so the rockwool takes up less chlorine from the water. However, most growers have not the time or the patience for this method, and simply just allow the block to sink to the bottom making sure the rockwool is completely saturated before removing the block for the next step in this rockwool process.

Once you have completed the above, then you should either give the cube a light squeeze reducing the volume of water held in the block by approximately 10%, or by flicking a little water out by making a quick downwards motion with your hand whilst holding the block, again only looking to take the excess water from the cube by approximately 10%. This is done to make sure that when the seeds are planted that the rockwool has adequate air inside the block to support life so that the seed can breathe.

Rockwool, by its own nature, holds water like a sponge. Therefore, if you plant a seed into rockwool that is completely saturated, there's a good chance the seed will germinate, then the seed will promptly suffocate and drown due to the excessive amount of water and the sad realisation that it has no air to breathe. The result will be a mushy dead seed/seedling.

Once the rockwool has the right amount of solution in it, the seed can be planted. The best way of doing this is to make a hole approximately 5 mm beneath the surface using a toothpick, pipette or similar instrument, then with tweezers pick up the seed and place in the hole approximately 5 mm beneath the surface. Once the seed is in position, gently cover the hole over so the rockwool has totally cocooned the seed. It is advisable not to cover over the hole too tight as the seed has only an initial amount of energy to pop and get to the surface. If the hole is too tightly covered over then the seedling can get stuck, unable to break through the rockwool covering. The seeds do need to be cocooned, as this creates maximum humidity around the seedling and also ensures the seed is in darkness.

Once planted, the seed can take from 12 hours to 12 days to germinate depending on environmental conditions, and the age and viability of the seed.

Seedlings require low to medium light levels; high frequency fluorescent tubes are all the rage at the moment offering excellent colour spectrum for the seeds and vegetative growth (blue spectrum) and are very inexpensive to run with no worrying heat output via the lamps.

The temperature of the environment should be maintained at or around 23°C with the humidity level around 70-90%. If using 55 watt twin high frequency fluorescent lighting, then these tubes should be one to two inches from the canopy of the propagator.

It is strongly advised to germinate seeds inside a propagator, which you can obtain from any garden centre or grow shop. For the first week of this germination process, the air vents on the top of the propagator should remain shut to maximise humidity once again. Seedlings like very high levels of humidity to get them germinated and to get them to establish without any shock or drama. Once seeds begin to grow and start to develop proper leaves, the air vents can be opened. But gently does it. To start with, only open slightly and as the seeds get used to this new humidity level then open again slightly more, and again allow the seedling to adjust, then open fully.

As the seedling develops, roots will begin to appear through the one inch cube. It is a good idea to then transplant the one inch cubes into the three inch cubes. You should follow the exact same procedure that you use when preparing the one inch cubes. Then, as the plant establishes its roots into the three inch cubes, it is time to think about hardening off or further propagation in the hydroponics system.

Jiffy-7s

Another very popular method of seed germination is the use of Jiffy-7s. These are great for beginners and require the use of no technical equipment. Simply add water and watch them expand. These nifty little numbers are in fact compacted discs of compressed peat, wrapped-packed in a disc-like pellet. By submerging these Jiffy-7s in water, they begin to expand by absorbing water and they grow to around 4 to 5 times their original size. Normally, it takes approximately 5 minutes to complete this process. Simply allow to drain for 5 minutes, then make your hole and plant your seed, then cover your seed over as you would if using rockwool.

These are good, however, most hydroponics systems require the seeds to be rooted through a three inch cube and as the jiffy only represents the size of a one inch cube, the grower cannot avoid the process that the seeds have to be transplanted to the three inch cube once the seedling has grown through the Jiffy-7. So in effect, they only temporarily avoid the inevitable, and as soon as they are big enough, you will need to do the all the preparation for the three inch rockwool cubes.

It is good practice to harden off the seedlings before transplanting into a hydroponics system. This process is a little time consuming but very worthwhile. Hardening off is a process of getting a plant adjusted slowly from one particular environment to another. In this case, from a propagator with diffused light and constant temperature and humidity, to the conditions outside of a propagator which will be slightly higher light levels, definitely lower humidity and possible fluctuations of temperature. This process of hardening off has already started through the opening of the air vents as the real leaves develop on the seedlings.

As explained above, once your seeds have established a good root system through the rockwool, you can now break your seedling into a new environment. This is done slowly and with patience. First, it is advisable to remove the propagator lid for a small period of time: 1-2 hours as an example for the first day. Then replace, wait till the following day, then remove the propagator lid for 4 hours, then replace and wait till the next day, then remove lid for 6 hours, replace, so on and so forth. This greatly reduces shock and gently hardens the plant off so they are adjusted slowly to the conditions of the new environment. Once they are used to having the lid off for 8 hours or more, they are ready for transplanting into the system. It is advised to be still using the propagation light during the process on an 18-20 hour light cycle. Now your seedlings are hardened off, the next step is to transplant into the hydroponics system.

On that note, some hydroponics systems allow the process of further propagation within the system. This again is advisable as when transplanting into the new system the seedlings are then going to be subjected to much heavier lighting levels and reduced humidity with fluctuations of temperature. Further propagation will reduce the shock of these new environmental factors. Same principle applies if you are going to do further propagation with the hydroponics system, break the plants in gently and harden them off slowly.

Cuttings

This section reflects many of the same principles as the seed propagation section, so there will be much repetition and overlap. However, for those simply looking up how to take clones or cuttings, we have included this section for reference purposes.

You have a number of different options available when taking cuttings. We will focus on each technique one at a time. Firstly, we will cover the different media you can use, then the actual method of taking a cutting.

Traditional Bare Rooting Method

This is a very simple way of getting your cuttings to take. Simply use disposable plastic cups, similar to the cups used in drinking water machines. Fill a cup with approximately 1-3 cm of water. Use normal tap water or bottled mineral water. Place your cutting in the cup using the side of the cup container as the support and rest the freshly cut stem under the water. Then place the entire cup with the cutting inside the propagator with air vents shut for maximum humidity.

Check daily never letting the cup's water level drop below 5 mm of the original water level. After 7-14 days, the cuttings should have sprouted roots; these are bare roots and can now be planted into a growing medium. It is still advisable to propagate the cutting in its new medium until it has established a good viable root system. This method is good for beginners who struggle to get results with other methods, which we will expand on shortly. It is vital that if the water shows any sign of stagnation, that the water in the bottom of the cup should be changed as soon as possible, and regularly, if this does occur.

The only drawback with this method is the cutting can suffer massive transplant shock and even possible damage to the roots when transplanting into the new medium. It is very easy to damage the cutting when actually picking up the cutting with fingers or tweezers. However, when cuttings won't take in the other methods mentioned below, then this is your best and last resort. Even the most competent growers have found, using rockwool as an example, that for all their experience and previous success, that somehow this time, the cuttings won't take in the rockwool substrate. So, the growers resorted to the method mentioned above and within 7-14 days, all the cuttings had rooted, fully ready for transplanting back into the rockwool for further propagation.

Rockwool

Modern rockwool methods are regarded as the most professional way of taking cuttings. This method is becoming more and more the norm used by nearly all hydroponicists. It involves using pre-soaked rockwool substrate; this is not normal rockwool but a specific one developed particularly for horticulture. The most popular brand of horticultural rockwool is Grodan.

The process is simple but ideally requires using specialised hydroponic electronic equipment. A digital CF meter and a digital pH meter is advised, available from all good grow shops. Briefly, the CF meter measures the conductivity factor of dissolved salts in the water. In English, this gadget measures how strong or how weak your nutrient mix is, giving you a yardstick to understand if you are possibly over or under feeding the plants. The pH meter is self-explanatory, it simply measures the current pH level of your water or nutrient mix. This again is invaluable, as this will give you another well needed yardstick in understanding and maintaining the right pH level for your plants.

Firstly, you have to make a stock solution for your rockwool to be pre-soaked in before you take your cuttings. Using normal tap water, simply fill up a bucket or container. The next step would be to add some weak nutrient so that the rockwool has food in it, so when the cuttings start to root they can get nourished quite easily.

If using normal 2 pack hydroponics grow nutrients, it is advised to start with only making this nutrient solution one quarter strength from the directions given to you on the side of the bottle. As an example, Canna nutrients are possibly the best selling nutrient available. These nutrients are stated to be diluted at 20 ml of A and 20 ml of B per 5 litres of water. So if you were to mix this nutrient for cuttings, you are advised to knock it up at one quarter that strength which would be 5 ml of A and 5 ml of B per 5 litres of water.

At this point the CF meter comes into its own as this little beauty can now verify, if or if not, you have the desired CF value for taking and establishing cuttings. A low CF value is required when establishing cuttings, it is recommended that a CF value of between 8-12 is reached.

If you find that the CF level is too high, then simply add water to dilute it down until you get to the right level. If you find that the CF level is too low, then add equal amounts of A and B solutions until you have reached the right level.

Please note nutrient solutions take time to mix and dissolve into the water, so don't be hasty, the nutrient might just be taking its time to mix and dissolve. Once you have established that you have the right CF level then it's time to balance and buffer the pH of the solution.

It is always advisable that the nutrient level is balanced first, as the nutrients can affect the pH level of the water you are adding it to. If you balance the pH first you might have to balance it again after adding the nutrient. This can be time consuming and is also not recommended for the plants, as the more pH up or down that you use, the worse off the plants will be, as both pH up and pH down attack vital food chains that are in the nutrient, thereby making it unavailable to the plant when it needs it. pH down is normally phosphoric acid and pH up is normally potassium hydroxide. In recent years a new pH down has come on to the market which is a blend of nitric based acids. This is apparently a food friendly based acid which will not damage any food chains like iron, zinc or calcium. However, if normal pH up and down are used correctly, then the nutrients or the plants will not mind it at all, they will actually greatly benefit from the process of having the pH maintained at the correct levels.

Again, another word of warning; never, ever mix pH up and pH down together directly as a concentrate as these two chemicals are highly aggressive and react violently when mixed together.

Once you have mixed the nutrient and are happy that you have it at the right level, then it's time to balance or buffer the pH. The rockwool likes to be pre-soaked in a nutrient solution with a pH level of or around 5.5. This is the ideal pH level for rockwool to be pre-soaked in. As the rockwool is slightly alkaline you need to make the pH solution more acidic to counteract the consistency of the rockwool. It is ideal if the cuttings are grown in a neutral pH level, and when reducing your pH level to 5.5 you are actually creating a neutral environment within the rockwool with the low pH counteracting the naturally high pH of the rockwool medium.

After adding your nutrient to the solution and adjusting it to the right level, it's time to test for your pH level. If your meter reads high you need to add pH down to it and if too low then you will need pH up. Both pH down and pH up are extremely strong and very aggressive so handle with care and only administer a tiny amount at a time to your solution, mix your solution well and allow the pH up or down to dissolve, then take another reading using your pH meter. If you find

the value is still too high or too low, then re-administer another tiny amount of your pH up or down.

If you find that you administer too much pH up or down then it is best practice to throw your solution away and start again, i.e. if you are living in a hard water area and you are adding pH down to your solution, if you add too much pH down into your solution, then you are best advised to throw it away and start again rather than trying to get the level back up with pH up. This is due to the fact that pH up and down are aggressive and when administered to the solution they will attack your nutrient. Adding pH up to counteract the pH down is obviously bad practice as you have administered too much in the first place, then you are administering more which does your nutrients, and for that matter, your plants no favours. Same applies if you are in a soft water area and you are administering pH up to your solution and again you administer too much pH up, then you are again advised to throw that solution away rather than to try and balance it out with pH down. Less is always best in the world of hydroponics.

Once your solution has been buffered with the right level of nutrient and balanced at the right pH level, it is now time to soak your rockwool blocks. Some growers allow the blocks to soak overnight and in practice is the right procedure when using tap water, as this allows the chlorine to rise to the top of the water so the rockwool takes up less chlorine from the water. However, most growers have not the time or the patience for this method and simply just allow the block to sink to the bottom making sure the rockwool is completely saturated, before they then remove the block for the next step in this rockwool process.

Once you have completed the above, then you should either give the cube a light squeeze reducing the volume of water held in the block by approximately 10%, or by flicking a little water out by making a quick downwards motion with your hand whilst holding the block, again only looking to take the excess water from the cube by approximately 10%. This is done to make sure that when the cuttings are planted, that the rockwool has adequate air inside the block to support life so that the cutting can breathe.

Rockwool, by its own nature, holds water like a sponge. Therefore, if you take a cutting in rockwool that is completely saturated, there's a good chance the cutting will take, but then the cutting will promptly suffocate and drown due to the excessive amount of water and the sad realisation that it has no air to breathe. The result will be a dampened off cutting rotted at the base of the stem.

Once the rockwool has the right amount of solution in it, the cutting can be planted. The best way of doing this is to make a hole approximately 10 - 20 mm beneath the surface using a toothpick, pipette or similar instrument, then with a pipette administer 1-2 ml of a good rooting gel compound into the hole then place the cutting in the hole approximately 10-20 mm beneath the surface. Once the cutting is in position, gently push the rockwool up tight to the cutting so the rockwool has totally sealed in the cutting. It is advisable not to push the rockwool over the hole too tight, as you do not want to inflict any damage to the cutting; the rockwool should be tight enough to offer good support for the stem to stand upright. The cuttings do need to be cocooned in, as this creates maximum humidity around the stem that will develop roots, but also ensures the roots are kept in darkness.

Some growers do not make a hole at all but simply dip the cutting straight into the rooting compound then push the cutting directly in the pre-treated rockwool. This method is the most common due to the fact it is easier to do this, however, the rooting compound normally gets rubbed off the bottom of the cutting and only the side of the stem of the cutting then benefits from this rooting compound. Also, the air available around the bottom of the cutting is reduced when using the "push it straight in" method.

Once planted, the cuttings can take from 7-14 days to take depending on environmental conditions, and the age and viability of the mothers.

Cuttings require low to medium light levels; high frequency fluorescent tubes are all the rage at the moment offering excellent colour spectrum for the cuttings and vegetative growth (blue spectrum) and are very inexpensive to run with no worrying heat output via the lamps.

The temperature of the environment should be maintained at or around 24°C with the humidity level around 70-90%. If using 55 watt twin high frequency fluorescent lighting then these tubes should be one to two inches from the canopy of the propagator.

It is strongly advised to strike your cuttings inside a propagator, which you can obtain from any garden centre or grow shop. For the first week of this rooting process, the air vents on the top of the propagator should remain shut to maximise humidity once again. Cuttings like very high levels of humidity to get them to strike and to get them to establish without any shock or drama. Once the cuttings begin to grow and start to develop proper roots and new shoots, the air vents can be opened. But gently does it. To start with, only open slightly and as the cuttings get used to this new humidity level then open again slightly more; again allow the cuttings to adjust, then open fully.

As the cuttings develop, roots will begin to appear through the one inch cubes. It is advisable to then transplant the one inch cubes into the three inch cubes. You should follow the exact procedure that you use when preparing the one inch cubes. Then, as the plant establishes its roots into the three inch cubes, it is time to think about hardening off or further propagation in the hydroponics system.

Jiffy-7s

Another very popular method of taking cuttings is the use of Jiffy-7s. These are great for beginners and require the use of no technical equipment. Simply add water and watch them expand. These nifty little numbers are in fact compacted discs of compressed peat, wrapped-packed in a disc-like pellet. By submerging these Jiffy-7s in water, they begin to expand by absorbing water and they grow to around 4 to 5 times their original size. Normally, it takes approximately 5 minutes to complete this process. Simply allow to drain for 5 minutes then make your hole, add 1-2 ml of liquid rooting compound then place the cutting in the hole approximately 10-20 mm beneath the surface. Ensure the cutting is in position and the Jiffy-7 should offer good support for the stem to stand upright. The cuttings do need to be cocooned in, as this creates maximum humidity around the stem that will develop roots, but also ensures the roots are kept in darkness. Jiffy-7s are good, however, most hydroponics systems require the cuttings to be rooted through a three inch rockwool cube and as the jiffy only represents the size of a one inch cube, the grower cannot avoid the process that the cuttings have to be transplanted to the three inch cube once the cutting has grown through the Jiffy-7. So, in effect, they only temporarily avoid the inevitable and as soon as they are big enough, you will need to do all the preparation for the three inch rockwool cubes.

It is good practice to harden off the cuttings before transplanting into a hydroponics system. This process is a little time consuming but very worthwhile. Hardening off is a process of getting a plant adjusted slowly from one particular environment to another. In this case, from a propagator with diffused light and constant temperature and humidity, to the conditions outside of a propagator which will be slightly higher light levels, definitely lower humidity and possible fluctuations of temperature. This hardening off process has already started through the opening of the air vents as the real leaves develop on the cuttings.

As explained above, once your cuttings have established a good root system through the rockwool, you can now break your cuttings into a new environment. This is done slowly with patience. First, it is advisable to remove the propagator lid for a small period of time: 1-2 hours as an example, for the first day. Then replace, wait till the following day then remove the propagator lid for 4 hours, then replace and wait till the next day, then remove lid for 6 hours replace, so on and so forth. This greatly reduces shock and gently hardens the plant off so that they are slowly adjusted to the conditions of the new environment. Once they are used to having the lid off for 8 hours or more, they are ready for transplanting into the system. It is recommended to be still using the propagation light during the process.

Now your cuttings are hardened off, the next step is to transplant into the hydroponics system.

On that note, some hydroponics systems allow the process of further propagation within the system. This again is advisable as when transplanting into the new system, the cuttings are then going to be subjected to much heavier lighting levels and reduced humidity with fluctuations of temperature. Further propagation will reduce the shock of their new environmental factors. Same principle applies if you are going to do further propagation with the hydroponics system, break the plants in gently, and harden them off slowly.

Taking a Cutting

It is always good practice to take cuttings from your best stock plants. Or your best mothers. On this subject, there are two ways of continuously revolving your crops with cuttings.

You can keep mothers that are perpetually in a vegetative state, i.e. subjected to a lighting cycle of approximately 20 hours per day. These mothers are kept in a separate room from the main grow room with their own dedicated light and environmental controls. This is important as you do not want any light pollution interfering with the grow room, and the mother and cuttings room. These mothers are kept trimmed back and are constantly pinched out. Pinching out is the removal of the centre growth tip from the middle of the plant. This is the grow tip which leads the plant. Once a grow tip has been pinched out, then more grow tips will follow and these too should also be pinched out. The pinching out of the growth tips will result in shorter, stockier and much bushier plants which will in turn produce a plant with many sites that cuttings can be taken from.

The more pinched out the mother plant is, the more cuttings you will be able to take from her. It is always advised to never take more than 40% of the volume of the plant in cuttings i.e. never strip a mother bare as she will most likely suffer extreme shock and not recover from this episode. When only taking half of the cuttings that she can support, the mother will quite happily regenerate in no time.

The benefits of keeping mothers is that you always

have at your disposal, the ability to take cuttings as and when you like. It is very important when choosing a mother plant that the best plants are used as their genetic make-up will become the same genetic make-up of the cuttings which you are taking from her. The disadvantages with mothers are, they need constant maintenance and take up more space than a cloning room would.

You can also take cuttings from the existing crop of stock plants that you are about to flower or fruit. This process is good as you can take the cuttings from the most promising of the stock plants, and in turn, from these cuttings once matured, you can also take cuttings from the best of these plants, and so on and so forth, until basically you have managed to obtain the cream of the cream. These can then be turned into mothers. These cuttings again like the mothers, should be kept in a separate room from the main grow room with their own dedicated light and environmental controls. It is even possible to take cuttings from your flowering

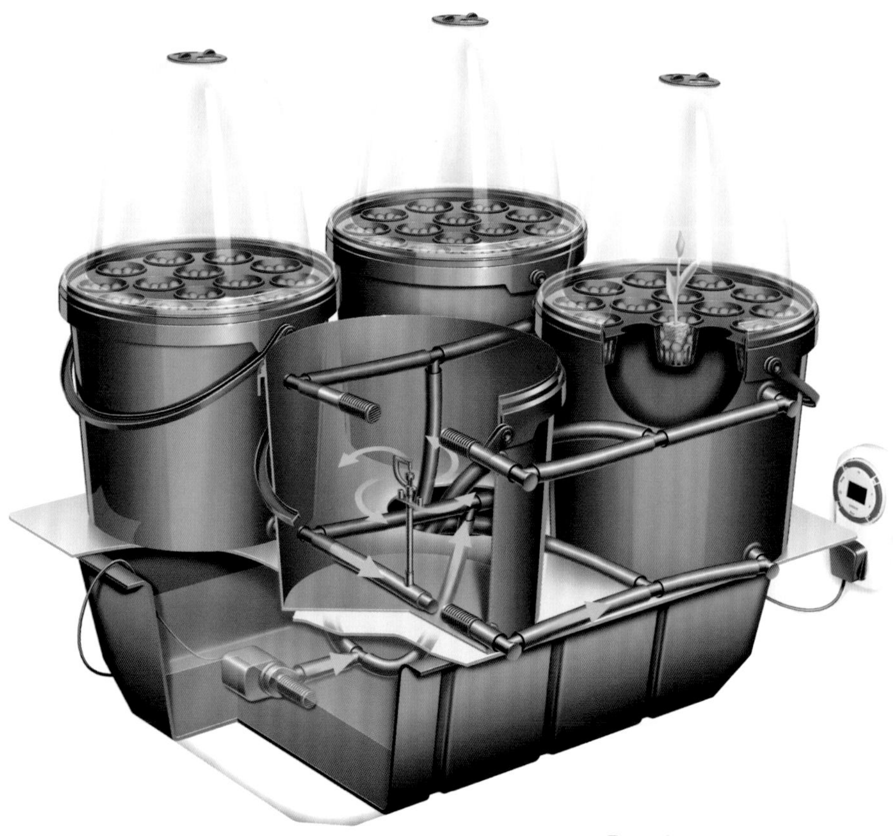

Example of an Automated Hydroponic Propagation System

plants as long as they have not been in flower too long. Cuttings can be taken from flowering plants up to 3 weeks into the 12 hour cycle. As long as these cuttings are then subjected back to a 20 hour daylight cycle, then they will flip from flowering back to vegetative growth, then once mature enough, can then be flowered.

This is slightly riskier than taking cuttings from a vegetative plant as sometimes they can be slow to take or refuse to be reverted back to the vegging cycle. If you are to pursue this method realistically, it is best to take your cuttings up to 2 weeks into the flowering cycle and this will give you a much better rate of return and no throw backs.

The benefits are that you can take the cream of the cream during each separate crop. The problem is that if the cuttings are taken two weeks into the flowering process, they have to wait almost 4-6 weeks before it is their turn to get into the grow room as you are waiting for the stock plants to finish. So, in effect, you have to keep your cuttings back for a prolonged period of time. Cuttings are normally ready in 2 weeks. So you could end up taking cuttings from your cuttings just to slow them down while waiting for your flowering plants to finish, and before long you have too many plants to know what to do with.

Cuttings Stage One - Preparation

Before attempting the process of taking the cuttings, it is well advised to get all the preparation done first and not during the process. The first stage of preparation before handling the equipment or the plants, is to disinfect or thoroughly clean your hands; there is good reason for this. You do not want to infect the plants with viruses or pathogens. Some growers have a tendency to handle rare or processed tobacco. Most growers do not realise that the majority of mass grown tobacco has embedded within it a virus known as tobacco mosaic virus, or TMV. This virus is very easily transferred from fingers to your favourite plants, especially when taking cuttings. If plants and cuttings get this infection then indeed your crops will not flourish as they should and the cuttings will have difficulties rooting, if at all. And if they do strike, the cuttings will be nothing to write home about. It is also important not to smoke in your grow rooms as the smoke too can carry this virus. And your fingers after smoking will have concentrations of this virus, which can be easily transferred onto the plants. Once you have cleaned or sterilised your hands, the following will be needed:

You will need a set of sharp sterilised scissors.

You will need a sterile scalpel or razor blade.

You will need a propagator, preferably heated.

You will need to prepare the growing medium.

You will need some rooting hormone gel.

You will need a sterilised bowl or cup half filled with plain water.

You will need a sterilised chopping or bread board.

Cuttings Stage Two - Choosing Your Mother or Mothers

It is advised to only take cuttings from the best specimens that you have, as the cuttings will be in fact clones of the original mothers and will show the same vigour and growth potential as the mothers. Some growers feed the mothers or stock plants on plain pH-adjusted water for a few days before taking the cuttings, as this tends to help the cuttings root quicker as the nutrient within the plants has become depleted. This makes the cuttings hungrier for nourishment and therefore encourages the cuttings to strike quicker to establish the roots so it can start to feed again.

Cuttings Stage Three – Taking Your Cuttings

More mature growing tips root easier than young new shoots. Also on that note, growing tips from the side and bottom of the plants root better than growing tips from the very top of the canopy.

With your clean hands and sterilised scissors, cut a growing stem of a branch approximately 3-4 internodes down from the growing tip. An internode is the junction on the stems where leaf and stem growth occur. Immediately place this long cut into the bowl or cup of water. Then, continue this process till you have the number of desired cuttings that you want. It is advisable to take a few more cuttings than required in case some do not strike.

Take your bowl of long branch cuttings to your sterilised chopping or bread board. Take only one cutting out from the bowl at a time and place on the board. Get your sterilised scalpel and remove the lower leaves from the third and fourth internodes, so all you are left with is a long stem

with two good sized leaves at the top and a young set of developing leaves at the growing tip. Depending on the size of the cutting you want and the size of the internode spacing, take a long elongated cut through the second or third internode. It is advisable to take a 45° angle cut

directly through the internodal junction as this is the area where most of the energy is within the cutting. Tests have shown that cuttings taken at the internodal point, root quicker, stronger and healthier than cuts made in between the internode joints.

Cuttings Stage Four - Planting

Immediately after cutting through the internode, dip your cutting directly into the rooting gel, administer a couple of drops of rooting gel directly inside the hole within the rockwool block, then plant the cutting straight into your growing medium which has already been prepared as explained above. Place your cutting approximately 1-2 cm down into your growing medium. The reason this is done immediately, is that when taking the initial cutting and placing it directly into water, it stops any air uptake through the open stem which can block the cell structures and actually has the effect of suffocating the cutting. This again is the reason that the next stage of the cutting process is taken using a sterile scalpel. The cut a scalpel gives on a stem compared to scissors is one that is cleaner, sharper and actually does less damage to the cell structure. If you were to take cuttings using

your scissors, this will result in the stem edges becoming blunted, rough and even pinched which will have the same effect as the air blocking or stifling the stem cell structure.

Cuttings Stage Five - Propagation

Place your cutting directly into your preferably heated propagator, put the lid on tight and keep your air vents shut. If you have a variable heated propagator, maintain the heating level at around 75°F. If during the lighting cycle the heat in the propagator exceeds 75°F, turn the propagator off during the lighting period and on during the night period so you do not get a major differential between day and night time temperatures. It is advised to subject the cuttings to approximately 18-20 hours of light but definitely not 24 hours, as most other authors tend to suggest. Cuttings need to rest just like the rest of nature. After a sleep as it were, they come back the next morning with much more vigour and strength for the day ahead. Water the cutting sparingly but never allow the medium to dry out, but also never allow the medium to become waterlogged. The medium the cuttings are in, and the plants themselves, will tell you if you are over or under watering. Keep humidity levels high at around 70-90% until new roots and leaves appear.

Cuttings Stage Six – Transplanting for Further Propagation

As the cuttings strike, roots emerge. Wait until roots are showing through the medium before transplanting to the bigger medium. It is recommended for beginners to transplant the newly rooted cutting directly into a three inch rockwool cube, then to place this cube and cutting back into the propagator for further propagation. Although your cutting can be transplanted out directly from your one inch, it is safer to get the cutting to root into the three inch first. As the cutting takes to the three inch cube, it is time to harden off and transplant to your system.

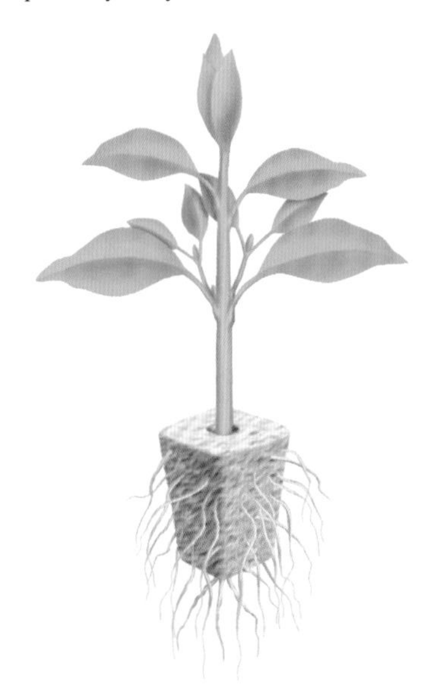

Chapter 4

Hydroponic Systems

Before we launch into the deep end, it was thought best to explain the differing hydroponics techniques that exist, and explain the pros and cons on each differing technique.

Nutrient Film Technique aka NFT

This was the original pioneering hydroponics technique. Although dated, it is still very popular among indoor horticulturists, mainly due to the inexpensive costs of setting up a NFT system and its simplicity.

The principle is very easy to grasp; the plants are grown in a constant flow of nutrient enriched water. The water is spread out so as to flow in approximately 1-3 mm of depth over a flat surface. This creates a film of water which flows over the root system of the plant. This is not a rapid flow but enough of a flow that the water is in constant motion. Water is fed to the table via a submersible pump from the top end of the table. The table is positioned at a slight angle to help the flow of the nutrient down the table. As the water is pumped in at one end of the table, it slowly makes its way to the bottom of the table which then returns back to the reservoir in which the pump is submerged. So you get the constant exchange of the water in the reservoir being pumped from one end of the table, then returning to the reservoir via the other end of the table. The film of nutrient should always be maintained at or around 1-3 mm of water. The roots of the plant should grow below and above the water's surface and that is why the film should be constant, allowing the water roots to develop below the water's surface and also allowing the air roots to grow above the water's surface.

The drawback of this system is that as the roots are constantly submerged in a film of water, this prohibits aeration to the rootball, which in turn prohibits performance. To get over this problem, some NFT growers put the pumps on cycles, effectively flooding and draining their NFT system. Other growers put air stones in the water reservoir and even under their plants on the NFT tables. Most NFT growers administer H_2O_2 to their reservoirs but at a very diluted ratio, however, this really needs to be done on a daily basis as diluted H_2O_2 breaks down very rapidly and over the course of 24 hours has completely dissolved its active ingredients. In using H_2O_2 in a daily capacity, this prohibits the use of organic growth promoters and other products that reduce the possibility of bacterial break out like pythium.

The main disadvantage with NFT systems, especially in a grow room environment, is the fact that pump failure is likely to strike at some point. The reason this tends to happen is that NFT systems are packaged with small low flow rate pumps; cheap springs to mind but this is not technically fair. The plants only need a small delivery of water at a constant rate and the small pumps are all that can be used on small NFT systems. Now as the pump is perpetually on, the pump sees a lot of action over the course of its life. This coupled with the fact that you are then adding dissolved salts into the reservoir and in turn you are possibly in a hard water area, you get precipitation of salts and calcium that build up on and around the impeller of the pump. Once this impeller begins to attract precipitation, it is not long until it either gives up spinning completely or that is does not deliver enough water to satisfy the plants' needs, resulting in crop failure. Pump failure can be overcome through regular cleaning and maintenance of the pump or indeed regular replacements of the pumps, and as mentioned earlier, these are very inexpensive pumps and therefore, can be regularly replaced without financial worry.

Another drawback with this technique is because the roots are constantly submerged in water, the plants are a lot more prone to bacterial diseases like pythium. Again, this can be overcome by regular dumping of the nutrient reservoir and adding products to the nutrient solution that have active ingredients that minimise the threat of root rot and moulds.

The last downside is that heavy yielding plants tend to fall over in a NFT system. This is due to the fact that the roots grow out flat and long, giving the plants no stability. As they grow older and bigger, you will need to support the fruits or flowers, otherwise, they simply topple over. Supporting them is easy using yo-yos, string, canes or some growers use a scroge. This is simply netting stretched out over the growing area of the plants. The plants grow up through this netting which in turn helps support them.

All of the above to one side, these systems are very productive and are an excellent and inexpensive teaching aid to the principles of hydroponics. Also, with this beautiful innovation, the world of hydroponics might not be with us as this technique was the first that was adopted and used by many growers all over the planet, paving the way for our very own hydroponics revolution. One has to take one's hat off to the British inventor that pioneered this technique. I mean, what made someone think "I know, let's grow plants in a soilless medium using nothing but a film of nutrient to do it in". ☺ Off the wall you could say!

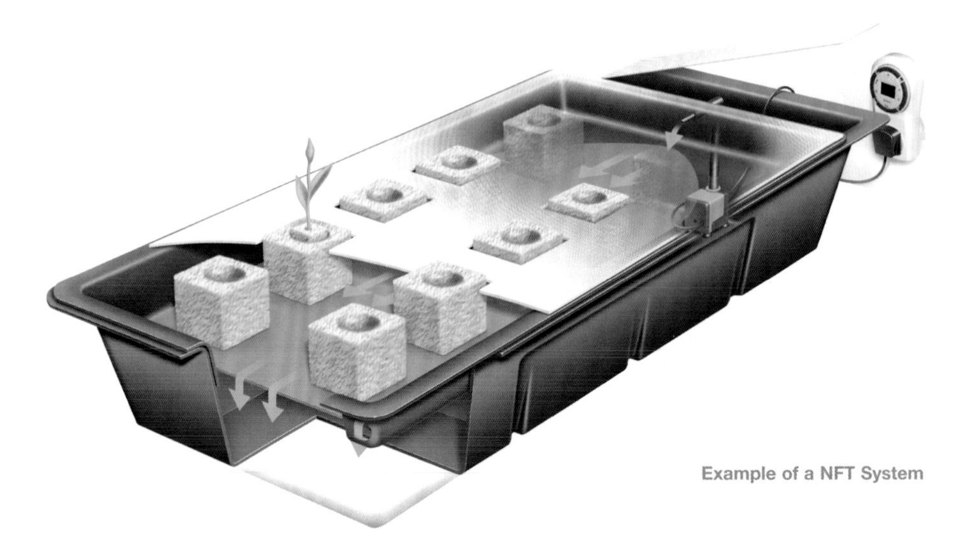

Example of a NFT System

Drip Irrigation Systems

The Dutch, who grow everything with this, have mastered this technique. Nor do they just grow everything, they grow on a scale unprecedented by any other nation. The plants are in a rockwool cube, then grown on in a rockwool slab. The plants are individually fed using drippers. These drip emitters are designed to deliver, at a set rate, a preset volume of water per hour. Each dripper is wired to an infrastructure of tubes and delivery pipes, which is fed by one master pump. Most commercial systems are what are known as high pressure drip systems, and most domestic systems are known as low pressure drip systems. These commercial high pressure drip systems are typically run-to-waste systems. This is when the nutrient is bled off and after dripping through the rockwool slab, then simply allowed to run down the drain i.e. to waste. This ensures these plants get the exact and maximum nutritional value from the nutrient solution and also cuts back on the possibility of bacteria or fungal problems like pythium. Most low pressure drip systems are recycling or re-circulating systems where the nutrient returns to the reservoir, then gets pumped back to the plants, then to the reservoir and so on and so forth.

These systems are relatively cheap compared to others and easily built once you get your head round the spaghetti of pipes, tubes and fittings. Drip systems are also very versatile and can be made in many shapes and formats allowing you a more modular design for your grow room. The running costs of these systems are not cheap, as after each crop you basically dispose of the rockwool slab and replace it with a new one. One 1 metre slab normally holds approximately three plants which can run up quite a bill if you have many plants. Also, if you are running to waste, then the cost of nutrient is very expensive indeed.

The main drawback with these systems is that the dripper can clog. Similar to the NFT system, if the flow of water stops, your plants will suffer. If you fail to notice that a dripper has stopped, the plant runs the risk of dying or at least losing its potential to give good yields. These systems need constant maintenance and upkeep and are not recommended for the beginner.

All drip irrigation systems are a little tricky to maintain, as you need to take constant pH and CF readings from your reservoir, from your run off and from your rockwool medium. This process is achieved by using a syringe to suck up a sample from inside the rockwool where the plants are growing, as the medium itself will hold a different pH and CF value to the reservoir and indeed, even the run off. Then armed with this info, you need to re-adjust your nutrient reservoir to compensate for what is happening inside the rockwool medium and the run off. Then, once this has been achieved, you then need to run the system again and do all those tests once more until you are happy that you have the right levels that you require. This process can easily need doing once a day. It is also advisable when growing in rockwool slab culture, to flush the salt out every two weeks with pH-adjusted plain water. This needs to be done, as rockwool tends to absorb unused salts which can build up, therefore, needs flushing through ever two weeks or so. It is also advised to do this flushing out process to expel any salt build-up that can concentrate in the drippers. Regular flushing can alleviate some of the maintenance of these systems. However, getting blocked drippers is part and parcel of this system so it is always advised to clean them regularly and also, to have ample spare drippers to swap when old drippers need cleaning.

In conclusion, this is a very productive hydroponics technique and has served the Dutch very well. The rockwool slab does offer a lot more support than you would achieve if you were using a NFT system. It also is less prone to pump failure, and as the slabs absorb a lot of water you do have some breathing space if the pump or drippers fail. It is very detachable and modular allowing easy expansion or removal of the system. Low pressure drip systems are more prone to dripper failure compared to high pressure drip systems, however, for the Percy Throwers, the high pressure option is too expensive and industrial for a small indoor garden. Overall, high levels of maintenance are required on both types of drip systems. The pipe work and drip lines also need regular replacement to combat clogging and salt build-up.

Ventura Action Drip System

Another entirely different drip irrigation system is that which uses Ventura action to deliver the dripping effect. These are individual grow tubs or pots specially designed for the smaller gardener. The system consists of a large outer pot which acts as a small water reservoir. Inside this large pot is a shorter, smaller inner pot which holds the grow medium which is typically clay pebbles. This smaller pot sits inside the bigger pot but does not actually hit the water level of the bigger pot, which is the holder of the nutrient solution. In English, it is a pot within a pot, the smaller inner pot is where the plants grow, the bigger outer pot acts as the reservoir. A Ventura pipe is then placed through the upper grow pot and submerged under the water level of the outer pot. Air is pumped down the Ventura pipe which causes the water to be pushed up above the level of the top of the grow pot. This is then piped into a delivery tube with large holes punched into it. The tube runs completely round, completing a circle joining back up to the Ventura

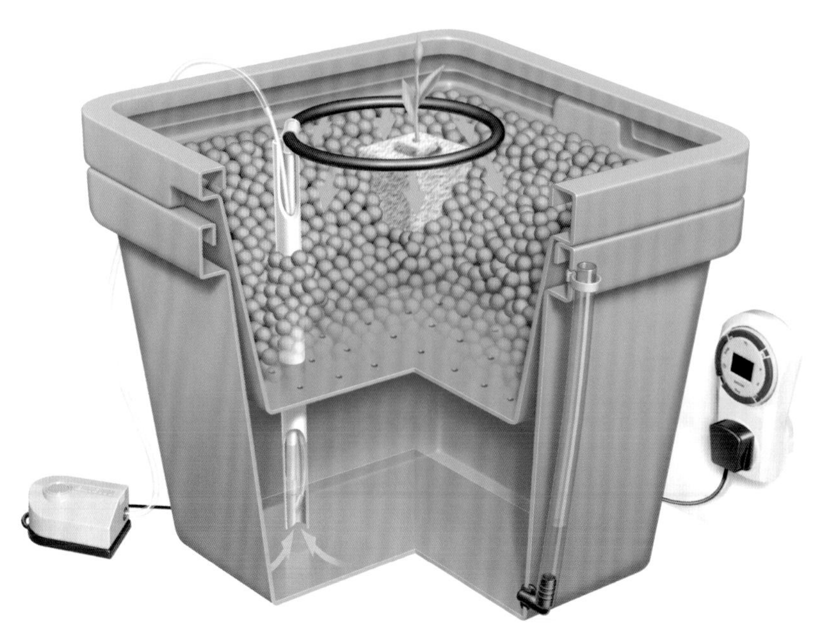

Example of a Ventura Drip System

pipe. Water pressurised via the air pump is then delivered through this drip ring which slowly but perpetually drips on to the clay pebble medium. The nutrient solution then drips entirely through the clay pebbles back to the outer reservoir, which in turn is then pumped back via the Ventura pipe to the top and delivered to the clay pebbles. If you like, it is a cross between a NFT system and a drip irrigation system but uses clay pebbles as the medium for the plants to grow into. Due to the fact that an air pump is used to deliver the water through the Ventura pipe, the nutrient solution delivered is highly aerated. Also, the constant dripping effect pulls air down through the clay pebble medium.

This system is typically only used for 1 to 3 plants or mothers; the reason for this is that to grow more, would require more pots. Each has its own individual reservoir. This reservoir, due to its small size, needs regular upkeep and to maintain lots of plants, you would need to maintain lots of reservoirs, so this would be too big a time consuming enterprise to undertake. Also, to adjust the inner reservoir, you have to lift out the smaller inner pot in which the plants are growing in; this again can be a costly exercise as it is very easy to damage the plants when lifting out and replacing this smaller inner pot. In recent years, a controller has been invented so you can link multiple individual systems together. But as the individual pots still have individual reservoirs passively connected to the controller, the controller fails to do its job in terms of pH and CF management i.e. you get differing pH and CF levels in each of the individual pots, but also in the controller itself, making it very hard to maintain precise control of your lovely crops. Due to the small size of the outer reservoir, the system will need daily maintenance to keep the reservoir topped up, and the pH and CF at the right level.

Algae is also very prone to develop in this system as the medium is fed from the top down, ensuring that the top of the medium is continually wet. This constantly wet medium being exposed to long periods of light, will always result in algae breakouts. As each system is packaged with its own individual air pump, the pumps after prolonged use can stop, having the same effect but not as frequently as the NFT system, resulting in a failed crop. On this note, the drip ring can also become blocked up with salt build-up and calcium deposits, so this too needs regular cleaning.

All of the above to one side, this is an ideal first system and is an inexpensive and valuable teaching aid to the hydroponicist. The system provides good aeration to the rootball and excellent support to your bigger yielding plants. It is good for mothers, but it is necessary to be very careful when removing the inner pots to top up and adjust the CF and pH of the outer reservoir.

Deep Water Culture aka DWC aka The Bubbler

This is the simplest method of hydroponics and is used mainly by growers who only wish to cultivate a minimal number of plants. It is a true domestic incarnation of hydroculture. This technique is cheap and works well, however, is high maintenance and not very easy to manage.

The system consists of a bucket with a lid, large net pot, clay pebbles, large air pump, airline, and a large round air stone. The air stone is placed at the bottom of the bucket with the airline running to the air pump on the outside of the bucket. The bucket is filled to approximately two thirds to three quarters full with water. The lid is placed on top of the bucket which holds a net pot, which in turn holds the clay pebble growing medium.

The system relies upon an air pump and an air stone to bubble air through the nutrient solution to mix it, but also in the beginning to generate spray, so as to get the clay pebble medium wet in order to establish the roots through the net pot. Consequently, the water has to be high enough to saturate the clay pebble medium by means of getting it wet or even moist via the actions of the bursting air bubbles. Once the plant has established a good root system, then it is advisable to lower the level of water in the bucket so that some of the roots can be allowed to hang in the air between the net pot and the water level. The majority of the roots live in the aerated deep water, 24 hours a day. With 24 hours a day in mind, as the roots are constantly submerged, it is crucial that the air pump is on 24 hours a day. If the pump is allowed to be off for any length of time, the roots will suffer from being waterlogged and starved of oxygen.

You can only typically grow one plant per bucket. Any more and the roots that develop can engulf the air stone and if this happens, then again lack of aeration to the rootball will result in death. This system does not operate via a reservoir, as the plants live in the reservoir itself. As the bucket can only hold a very limited amount of water - 5 to 10 litres, and the plants are living in this small amount of water, the CF and pH of this water is in a state of constant flux and therefore in need of constant attention. Not only that, but larger plants can deplete 5 litres of water in a single day so you are also in constant danger of the system literally running out of water. To emphasise this a little better, as plants uptake water they do not necessarily uptake nutrients at the same rate and as this occurs, the plants might be drinking lots of water but not eating much food. The result is that in a matter of hours, as the water is depleted, the concentration of the nutrients in the bucket can reach toxic levels. So it is critical to always under feed your plants in this system.

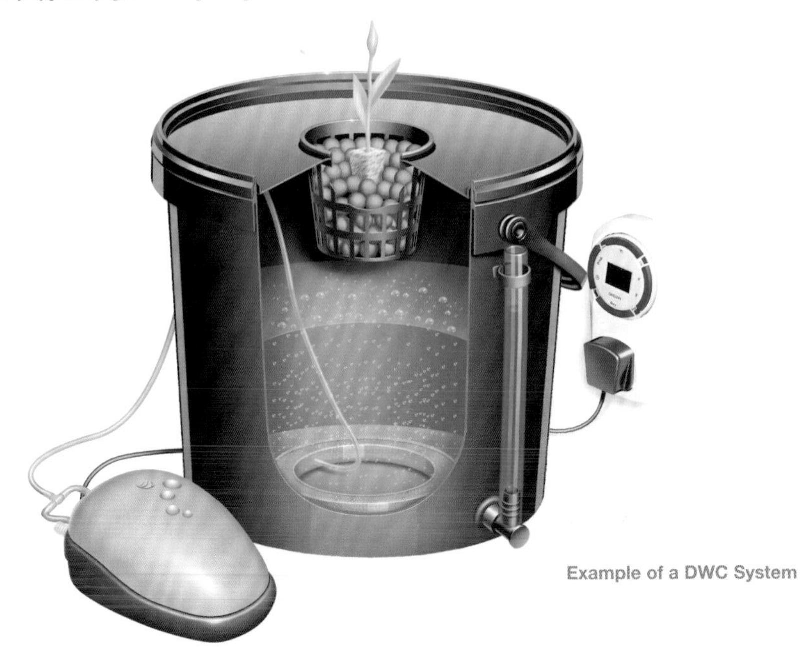

Example of a DWC System

With this in mind, the pH will also fluctuate as the water is depleted but the nutrients are not. So, you are advised to keep a constant check on monitoring and maintaining this system. Maintenance of this system also presents a headache because in order to change the water and check the pH and CF, you have to physically remove the lid and the plant from the growing chamber. To lift the lid with a small plant in it is not so much a worry, but to be constantly lifting the lid with a large plant located in it will do damage to the plant and the rootball of the plant. Bear in mind that you will have to do this on at least a daily basis, so the practicality of this technique is very questionable. If you were to maintain several of these buckets in one grow room, then you could be looking at a full-time maintenance job – not really what hydroponics is about! Large plants that are grown in this system will also need extra support due to the fact that the plant is grown in a minimal amount of medium, which will not support heavy yielders.

As this system's engine is an air pump and an air stone, then the bigger the air pump and the bigger the air stone, the better this system will perform.

Although the DWC has its drawbacks, these systems work very well and cost nothing to make or buy. A very inexpensive, productive and effective introduction to hydroponics, and a very good learning tool due to the amount of maintenance involved!

Flood and Drain aka Ebb and Flow Systems

This is a very straightforward and effective technique. It is basically a hybridisation of all the above techniques put together. The principle, as its name suggests, is simple but very impressive. These systems normally use clay pebbles as the substrate as this medium provides very good drainage and good retention of water, which over time will dry out. A flood and drain system typically works by using a timer and a submersible pump. The timer controls the flood and drain cycle of the system. Most flood and drain systems work via a bottom flood, which over a period of preset times, floods two thirds to three quarters of the growing medium. Then, once the flood cycle reaches the desired height, the pump stops and gravity then pulls the water back to the reservoir. Then, depending on the size and depth of the system, some time later the cycle is then repeated.

Conventional flood and drain systems are normally quite shallow i.e. approximately 5-10 cm deep. However, due to recent developments, we now have on the market deep flood and drain systems. These systems are approximately 50 cm deep. Original shallow flood and drain systems obviously take less time to flood, however, needs to be flooded more often over a 24 hour period. The new deep flood and drain systems take a longer period of time to flood, but require fewer floods during a 24 hour period. The reason for this is that the medium takes shorter and longer periods of time to dry before another flood is initiated i.e. the shallow flood and drain takes less time to flood and drain, however, the medium will dry out quicker. The deep flood and drain takes longer to flood and drain but as it has more medium in it, takes longer to dry out resulting in less floods per 24 hours than the conventional shallow flood and drain systems. The effect of a flood and drain system is similar to a piston in a cylinder of an engine. The raising of the water level during the flooding cycle pushes the old air out of the medium and therefore the rootball, then, when the flood stops and the drain starts, the suction caused due to the lowering level of the water, pulls new air into the medium and therefore the root system of the plants. The static period when the medium dries out again also pulls air into the roots. The result is superb aeration to the root system. The deep flood and drains provide substantially more aeration to the root system compared to the shallow flood and drains.

So to recap, the system works by flooding the system to almost the top of the growing medium. Then the system is allowed to drain, normally by gravity. After a period of time, normally once the medium is virtually dry, the system then repeats the cycle and so on and so forth for the duration of the crop. The flood and drain cycle is set at regular intervals during the 24 hour period and even set for the night cycles, as although the water and nutrients is not usable to the plants at night, it does exchange old air for fresh increasing the aeration around the roots which will prohibit bacterial and mould infections, but also keeps the water in motion which again will prohibit stagnation and bacterial problems. Flooding during the night period also increases the oxygen absorbed by the water, which again keeps the nutrient solution healthier and more usable to the plants.

Flood and drain systems are mostly recirculating systems and not run to waste. They typically need larger sized reservoirs compared to the other hydroponics techniques on the market. The benefit of this is the greater the size of the reservoir, the less maintenance it will require. The bigger the reservoir, the better pH stability you will have in the system and the greater control and buffering of the CF in the reservoir. This means less adjustments and less visits as the larger the reservoir, the more water the plants can uptake before you need to refill it. The one downside is that the larger the reservoir is, the more it weighs and this can present some problems for the upstairs gardener.

Before any more praise is said about these systems,

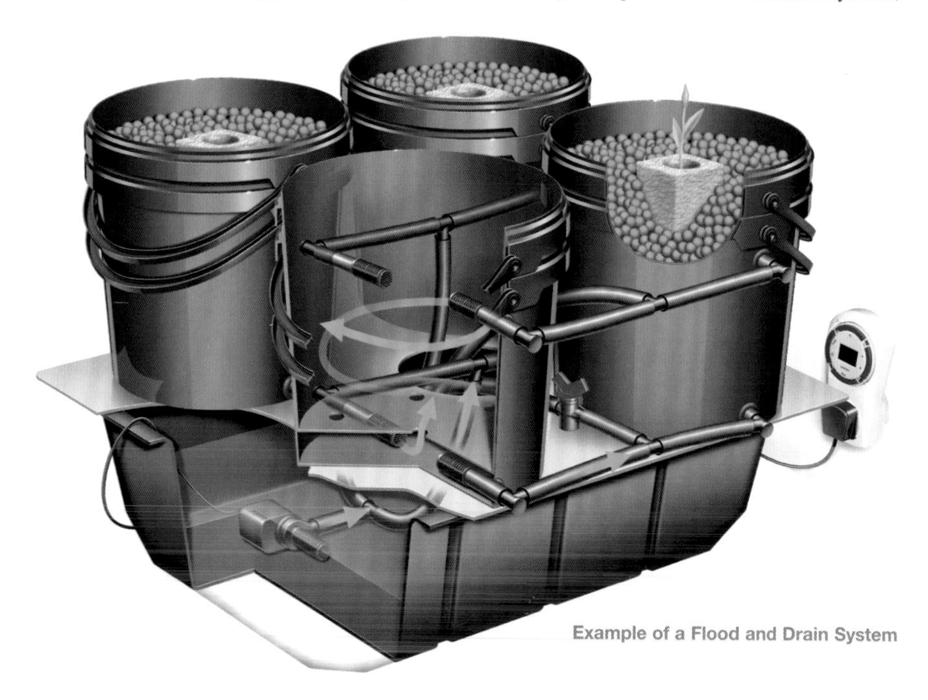

Example of a Flood and Drain System

please read what the horticultural press said upon their release:

Future Grow Magazine Excerpt

"These systems incorporate brand new evolutions in hydro farming innovation. Simply put, these are deep pod ebb and flood. Each pod has a large 12-15 litres of root space allowing for a longer dry period between floods optimising air to the root zone. Due to the depth of the pod (9-10 inches) and quality of the pump, the system takes 3-4 minutes to flood and 3-4 minutes to drain, resulting in a rapid flood and rapid drain. The water, due to its large volume coupled with its tall cylindrical pods, acts as a big piston pushing all the old air out then sucking new air in with approximately three times the pull compared to existing shallow depth flood and drains that are generally available, thereby getting considerably more air to the root zone. The system incorporates a couple of safety features, one being a shallow reservoir built into each pod so if a power cut strikes or the pump fails, your plants won't die. The Hydro Pod also has an overflow safety feature so if your pump gets stuck in the on position, then the whole system acts as a very deep trough NFT; no water spillage all over the floor to mop up. With the depth of grow pods the system incorporates excellent support for your bigger plants. Again, due to the size of root space, you can grow up to 3 plants in each pod. The Hydro Pod system can come built to measure, the systems can be built into any shape of room optimising the space available, the smallest being a two pod, the four pod, then an eight pod, then sixteen pod system, to whatever size or shape you wish. Each pod can be removed separately without disturbing the existing pods within the system, allowing you to start and finish plants at separate times or enabling the removal of diseased and unwanted plants without disturbing the other plants in the system. You can even turn the pods individually creating more even growth. We believe no such system has had so much thought and time put into its evolution which is still ongoing. Plans to develop an ebb and flow aero pod along the same principles are in the pipeline. If you're into HID light cultivation, the sixteen pod system can fit snugly into 1.5 m^2 ; that's 16-48 plants under one 600w or 1000w light." New Products: Hydro Pod Ebb and Floods from Future Grow Magazine

Flood and drain systems are easy to install and are very user-friendly. The secret to their success is their simplicity. Ideal for the beginner or the professional. Provides an excellent foundation for expansion. The only drawback, and yes, there is one, is that as with any hydroponics technique that employs clay pebbles as its medium, the clay pebbles do need to be washed thoroughly before use and between crops. Apart from that, this technique is a real winner.

Aeroponics

The principle of true aeroponics works by creating a fine mist of nutrients sprayed inside a tube or container in which the roots of the plants are suspended. So, in effect, the roots are fundamentally hanging in the air supported by the bare minimum of medium, i.e. a net pot with either clay pebbles or rockwool. The roots are then constantly sprayed with a fine mist via a very powerful pump. In principle, this is the definitive technique as the mist itself absorbs very high levels of oxygen plus the roots are hanging in the air, resulting in the absolute maximum aeration to the root zone. However, in practice, the majority of aeroponics techniques fail and fail miserably. The reasons for this are that the basic principle of making mists which have dissolved salts (nutrients) in it is flawed. Salts precipitate, this cannot be avoided, so over a short period of time salts build up on the misters restricting the flow of mist to the roots. If left unchecked, then total blockage occurs. Again, this effect is amplified if you are situated in a hard water area, as much of the UK is. The calcification of the water accelerates the blocking of these misters. The end product is total breakdown or constant daily management of the system, sometimes even hourly!

So, in theory, as a drawing and an idea, it's unbeatable, but in practice as a workable system, the conclusion is failure. If this was the first hydroponics technique that was invented, we would possibly not have hydroponics as we do today.

However, as with all good ideas, the simplification of them normally results in success and not failure. So that is what was done. The aeroponics pod system, instead of using a mister, uses a spinning high frequency sprayer. The sprayer delivers a constant fine spray direct to the rootball of the plants. As the sprayer is spinning at a high frequency, the spinning motion stops any precipitation of salts or calcification of the water on the sprayer, resulting in no blockages. Also, through its own action, it uptakes lots of oxygen in the process; ok not as much as mist would but compared to most hydroponics systems, it's leaps and bounds ahead of them. The roots hang in the

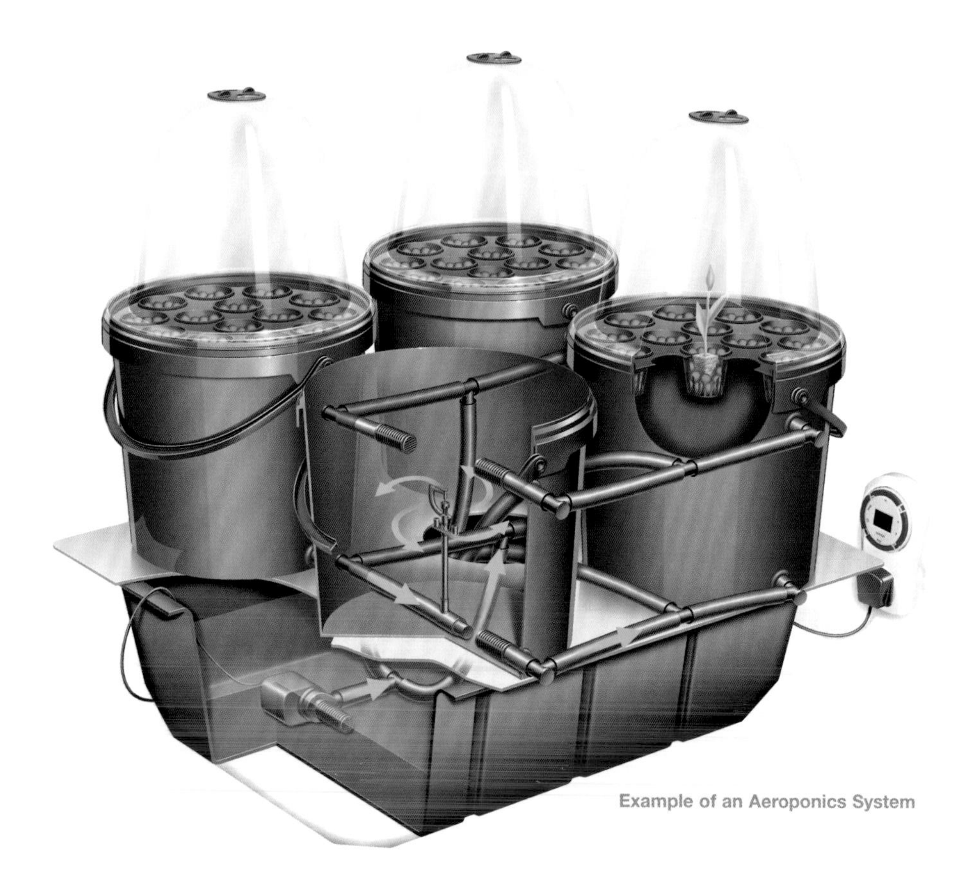

Example of an Aeroponics System

air, which again provides great aeration to the rootball. This system also incorporates all the safety features and flexibility of the original pod system.

This system offers good versatility as the lids are interchangeable. These systems can be used as an automated propagation unit, which can allow you to propagate up to 8 plants per pod! Or you can use it to propagate and grow to full maturity, 1-3 plants per pod.

Hydroponics
Indoor Horticulture

Chapter 5

Nutrients

Nutrient solutions and powders are a minefield of different and almost confusing variations, based however, on the same theme. It seems that the nutrient manufacturers want to confuse you into buying every possible nutrient, vitamin, hormone and a whole host of other additives to get you to spend as much as possible, in as many ways as possible, so your precious plants can be happier than they already are. To tell the truth and nothing but the truth, nutrient solutions are no more or less than glorified water, in which the most expensive part is the bottle itself!

However, the above to one side, these nutrients are absolutely critical for the nurturing, development and fulfilment of your plants' potential. No nutrients equals no worthy growth. To clarify, certain combinations of nutrients are an absolute necessity for your plants, however, the nutrients manufacturers have exploited this fact from every angle possible! Nutrients are simply differing variants of N-P-K. These nutrients are the required elements that plants uptake in order to grow. The plants absorb carbon, oxygen and hydrogen from the air and water; the other elements required to maintain growth are nutrients and light.

Nutrients can be subdivided into two groups: macronutrients and micronutrients. Macronutrients are the primary nutrients; micronutrients are the secondary nutrients, also known as trace elements. These nutrients can also be classified into a further two groups – stationary and mobile.

(N) nitrogen, (P) phosphorus, (K) potassium, (Mg) magnesium, and (Zn) zinc are mobile nutrients.

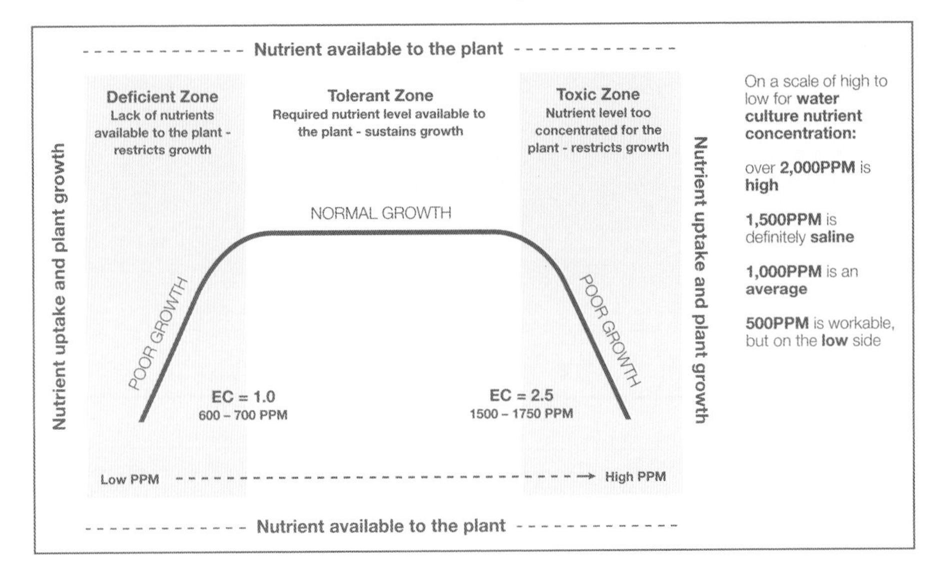

This means that they are able to translocate and re-translocate from one area of the plant to another.

(Ca) calcium, (B) boron, (Cl) chlorine, (Co) cobalt, (Cu) copper, (Fe) iron, (Mn) manganese, (Mb) molybdenum, (Si) silicon and (S) sulphur are stationary nutrients. This means that they are not able to re-translocate from one area of the plant to another.

An example of re-translocating nutrients would be (N) nitrogen stored in an older leaf system and in order to solve a deficiency in the younger leaves would be able to migrate to solve this deficiency. Deficiency symptoms and signs of re-translocating nutrients appear on the mature leaf systems first.

Now in the case of stationary nutrients, they remain static in their original place in the older leaf system. Deficiency symptoms and signs of stationary nutrients are first seen in the upper new leaves of the plant.

Macronutrients N-P-K

These elements are what the plants uptake the most. The percentage volumes of these elements are normally shown on the front or back of the bottled nutrient solutions. These nutrients are quite simply the base building blocks of all known life. Plants cannot thrive without them.

Example of Nutrients Required by Plants

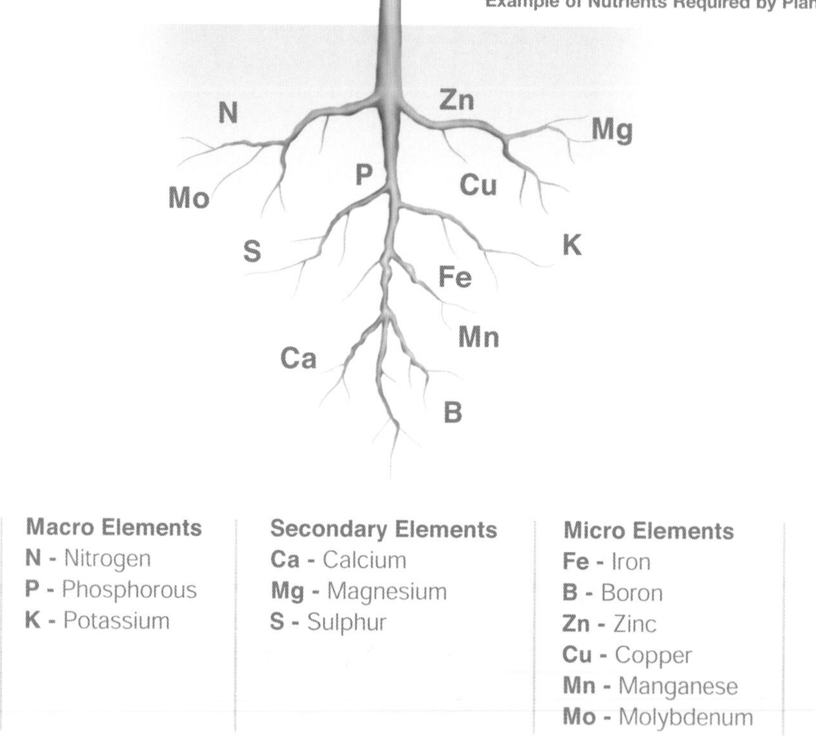

Macro Elements	Secondary Elements	Micro Elements
N - Nitrogen	**Ca** - Calcium	**Fe** - Iron
P - Phosphorous	**Mg** - Magnesium	**B** - Boron
K - Potassium	**S** - Sulphur	**Zn** - Zinc
		Cu - Copper
		Mn - Manganese
		Mo - Molybdenum

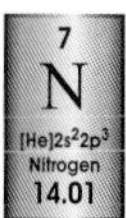

Nitrogen (N)

High energy plants require high levels of nitrogen, especially during the vegetative cycle of the plants' growth. Nitrogen acts as a regulator for the plant to manufacture proteins for new protoplasm essential for the cells growth. Nitrogen is also essential for the manufacturing of amino acids, nucleic acids, chlorophyll, alkaloids, and enzymes. This element is an important nutrient for the leaf and stem growth; it contributes towards the size and vigour of the plant's growth. Nitrogen is very active in immature buds or fruit, shoots and leaf systems. (NH4+) Ammonium nitrogen is the most readily available, however, do not overdo using this as it can easily burn your favourite plants. (N03-) the nitrate form of nitrogen has a much slower release than ammonium nitrate. Most hydroponics nutrients employ this type of nitrogen compound as a slower acting complement to the faster acting ammonium.

Deficiency

The most common nutrient deficiency is the lack of nitrogen. Telltale signs are that of slower growth. The lower canopy cannot manufacture chlorophyll and this encourages yellowing between veins while the veins themselves remain green. The yellowing migrates throughout the entire leaf, with the end result causing the leaf to die back and drop off. The stems of the plant and the leaf's underside turns a reddish-purple, however, this can also be a sign of a phosphorus deficiency. Nitrogen is a very active mobile nutrient that can dissipate into the environment quickly; this element does need to be added regularly to sustain fast growing plants.

Progression of this deficiency is that the older leaf systems yellow between the veins, then the older bottom leaves turn entirely yellow, then more and more leaves turn yellow and the badly affected leaves drop. Leaves and veins of the leaves' underside can also develop reddish purple stems. Progressively younger and younger leaves yellow between the veins, then all foliage yellows and finally leaf drop-off is severe.

Deficiency treatment is simple, by either using a complete quality hydroponics nutrient solution, or just a (N) nutrient solution. After an application of the above, the results should manifest themselves within 4-5 days. Sometimes bio-fertilisers or additives act as a catalyst to help the plant uptake nitrogen.

Toxicity

Strangely enough, the most common nutrient toxicity comes from nitrogen overdosing. This is evident by excessive lush foliage that is soft and susceptible to stressing, including fungal and insect attacks. The stems of the plants can become weak and may fold. Water uptake is restricted by the vascular transport tissues breaking down. Roots can develop very slowly with a tendency to brown and rot. Leaves in very bad cases can turn brown and copper, then dry and fall off. Flowers can end up small and sparse.

Progression of this toxicity is that you get exceptionally lush green foliage, weakening of the stems that possibly collapse, restricted root development, wispy flowers, browning of leaves that dry out and fall off.

Toxicity treatment is again very easy and simple. Flush your system with pH-adjusted water only for a day or two; then dump the reservoir of flushed water and replace with a new batch of nutrient solution but at a reduced CF than you were administrating before. Maintain this diet until the plant shows signs of recovery then gradually, over a couple of weeks, bring the nutrient level back up to the desired level.

Phosphorus (P)

High energy plants use high levels of phosphorus during the germination, young plant and flowering stages of the plants' growth. Bloom boost nutrient solutions designed for flowering have high levels of phosphorus in them.

Photosynthesis relies on phosphorus to provide a mechanism for the transference of energy within the plant. Phosphorus has many enzymes and proteins, and is associated with general vigour, resin and seed production. Concentrations of resins can be found in the growing tips of roots and shoots, as well as in the vascular tissue.

Deficiency

Phosphorus deficiency causes dwarfed or stunted growth, leaves are not as broad and are smaller, turn a greenish-blue and can develop blotches. Stems, leaf stems and the main veins turn a reddish-purple with the first signs on the leaf's underside, although the reddening of the stems and veins is not always very pronounced. The leaf tips of the more mature leaves turn dark and curl down. Severely affected leaves develop large blackish-purple dead blotches, then become bronzish-purple, dry out, shrivel, contort and drop off. Another sign of phosphorus deficiency is delayed flowering, fruits and buds are uniformly smaller and seed production poor, with increased vulnerability to fungal and insect attacks.

Progression of this deficiency is stunted and very slow growth rate, dark greenish-blue leaves often with blotches, smaller plants overall and as the blotches increase in size, the leaves turn bronzish-purple, mutate and drop off.

Deficiency treatment again is easy to administrate. Change the nutrient with a fresh batch of hydroponic nutrient solution and lower the pH in the hydroponics system to 5.5-6. After 3-5 days signs of improvement will occur.

Toxicity

Phosphorus toxicity may take weeks to manifest, especially if the excess phosphorus is buffered with a stable pH level. High energy plants use a lot of phosphorus throughout their entire lives and most high energy plants tolerate high levels of this element. However, overdoing it with phosphorus can interfere with calcium, copper, iron, magnesium and zinc uptake and stability. Phosphorus toxicity symptoms can manifest themselves as zinc, iron, magnesium, calcium and copper deficiencies.

Toxicity treatment again is very straightforward. Just dump the water in the reservoir and flush the system using pH-adjusted water only for a day or two. Then discard this flushed water and administrate a nutrient level with a lower CF level and maintain this diet until the symptoms retreat. Over the course of a couple of weeks, increase the CF to the desired level.

19
K
[Ar]4s1
Potassium
39.10

Potassium (K)

Potassium uptake occurs at all stages of the plant's growth cycles. Potassium also increases the plant's resistance to bacteria and mould, however, potassium is essential for the plant to manufacture and move sugars and starches, as well as in growth by cell division. This element also increases chlorophyll in the plant's foliage and helps regulate the opening stomata so that the plant can make good use of light, air and CO_2, and is vital for the accumulation and translocation of carbohydrates. In addition, potassium encourages strong, happy root growth and is connected with disease resistance and water uptake.

Deficiency

Potassium deficiency does not materialise straight away. The plants can initially appear happy and healthy, symptoms are that the older leaves, first at the tips, then the margins and then followed by the entire leaf, turn dark yellow and then die. Brittle or weak stems can arise and the plants become susceptible to disease. Potassium deficiency can cause the internal temperature of the foliage to climb and protein cells burn or degrade. Highest levels of evaporation normally occurs on the leaf edges and that is where burning can take place.

Progression of this deficiency is that the plant initially appears healthy with dark green foliage, leaves then lose lustre, branching is sometimes increased but branches are weak and scrawny, the leaf margins can then turn grey and progress to a rusty brown and curling and drying of the leaves. Older leaves then yellow accompanied by rust coloured blotches, leaves curl and rot sets in with the older leaves dropping; there is retarded flowering and diminished yield.

Treating a potassium deficiency is not complicated. Dump the old reservoir of nutrient. Replace with a fresh stock of hydroponics nutrient solution, lower pH to around 6-6.5 and after 3-4 days, signs of improvement will occur.

Toxicity

Potassium toxicity is occasional and is difficult to diagnose due to the fact that it is usually mixed with deficiency symptoms of other nutrients. Overdoing it with potassium will impair and slow the uptake of magnesium, manganese and even zinc and iron, but not always. So when symptoms of magnesium, manganese, zinc, and iron deficiencies appear, also look for a potassium toxic build-up.

Potassium toxicity treatment is very easy to administer. Simply dump the existing nutrient reservoir. Replace it with fresh pH-adjusted water, flush the hydroponics systems for a day or two, then replace the flushed nutrient with a fresh batch of nutrient solution but at a reduced CF level than before. Over the course of a week or two, increase the nutrient solution to the desired CF level.

Micronutrients

Micronutrients or secondary nutrients (magnesium, calcium and sulphur) are again used in large quantities by the plants. High energy plants are able to consume and process more secondary nutrients than a general purpose garden centre fertiliser is able to provide. Most high energy plant growers use a two part nutrient solution specifically for hydroponics cultivation. These two and even three part nutrients have all the macro, secondary and trace elements necessary for good plant growth.

Magnesium (Mg)

High energy plants use a lot of magnesium so deficiencies can be common. Magnesium can be found as the central atom in every chlorophyll molecule and is absolutely essential to the absorption of light energy. Magnesium helps in the utilisation of nutrients, and magnesium also neutralises acids and toxic compounds produced by the plant.

Deficiency

Magnesium deficiencies can be quite commonplace. The lower canopy of the plant develops yellow patches between darker green veins and brown rusty spots can appear on the leaf's margins, tips and between the veins as the deficiency progresses through the plant. The tips

of the brown leaves usually curl upwards before dying. Sometimes, the whole plant can discolour over a few weeks and if a severe deficiency is allowed to develop, the plant leaves will turn a whitish-yellow before going brown and finally dying.

In the case of a minor deficiency during the growth period of the plant's life, then little or no noticeable problems occur. But if minor deficiencies are left unchecked this can result in a diminished harvest as the flowering progresses through its cycle.

It is also worth noting that small root systems on plants have great difficulty absorbing enough magnesium to satisfy the heavy demands of the plants' growth.

Progression of this deficiency is that the plant looks healthy during the first 2-3 weeks, then the plant's lower to middle canopy develops yellow patches between darker, greener veins with possible rusty spots appearing on the leaf. Younger leaves on the plant look quite healthy. The tip ends of the leaves then turn brown curling upwards as the deficiency progresses, the brown rusty spots multiply and the yellowing of the veins increases. This progression starting at the bottom then progresses all the way through the plant, at which point, the younger leaves develop the spots and the yellowing. The leaves then dry and die.

Treating a magnesium deficiency is very straightforward. Firstly, make sure that the water in the system is not allowed to go below a temperature of 70°F as magnesium is very hard for the plant to uptake if the water temperature is too low or for that matter too high. So do not let the temperature of the nutrient solution get above 75°F. Make sure your nutrient solution is maintained at 6-6.5, or if in rockwool 5.5-6. If this has no effect, then remove old nutrients and mix up a fresh batch as outlined above. Within 3-7 days the deficiency will have passed.

Toxicity

Magnesium toxicity is occasional and is difficult to diagnose. If it is allowed to become very toxic, magnesium can develop conflict with other fertilisers and ions, typically calcium.

Magnesium toxicity treatment is very easy to address. Simply dump the existing nutrient reservoir. Replace this with fresh pH-adjusted water, then flush the hydroponics system for a day or two. Replace the flushed nutrient solution with a fresh hydroponics nutrient solution but administer it at a reduced CF level than you were originally on, pH-adjust the water to 6 and over, then over the course of a week or two, slowly increase the CF to the desired level.

Calcium (Ca)

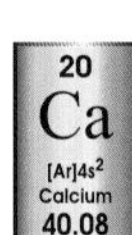

High energy plants require nearly as much of this stuff as other macronutrients. Calcium is responsible for the cell manufacturing and the growth of the plant. Calcium preserves and strengthens membrane permeability and cell integrity. High energy plants need some calcium at the growing tip of each individual root.

Deficiency

Calcium deficiencies are not usual in an indoor grow room. The deficiency starts with very dark green foliage and exceptionally slow growth. The youngest leaves are the first to show signs of deficiency. Strong calcium deficiency results in the new growing shoots developing yellowish to even purple hues and disfigurement before they shrivel and die. Flower development is inhibited, plants become stunted and the overall yield is compromised. The growing tips can show signs of deficiency if humidity is very high. If the humidity is very high, the stomata close and the transpiration of the plants stop. Calcium transported by this transpiration becomes immobile.

Treating a calcium deficiency is not hard. Dump the old nutrient solution and replace with new nutrients making sure the pH of the solution remains stable as large fluctuations can enhance the deficiency.

Progression of calcium deficiency is that first, young leaves turn a very dark green and the rate of growth becomes very slow. The new growing shoots and tips discolour, the new shoots then contort, mutate, shrivel and die. The flower and fruit development slows rapidly.

Toxicity

Calcium toxicity is hard to spot during the vegetative cycle of the plant's life. However, it does cause wilting. Toxic levels of calcium also exaggerate the deficiencies of potassium, magnesium, manganese and iron. Even if the nutrient is present they become unavailable. Too much calcium too soon in the early stages of development can also stunt the growth of the plant. Excessive amounts of calcium will also precipitate with sulphur in the nutrient solution. This can cause the nutrient solution to become cloudy. Once calcium and sulphur combine, they form gypsum which is a residue that settles to the bottom of the reservoir.

Sulphur (S)

Most nutrient solutions available contain sulphur and it is very rarely deficient. Most growers do not entertain pure elemental sulphur but prefer to use sulphur compounds such as magnesium sulphate. Nutrient solutions combined with sulphur mix better in water.

Sulphur is a base building block for lots of hormones, proteins and vitamins including vitamin B[1]. Sulphur also buffers the stability of the pH in the nutrient solution. Sulphur is used as an indispensable element in plant cell and seed development. Sulphur is fundamental for the plant's respiration and is active in the breaking down of fatty acids and the structure and metabolism of the plant. It is involved in the protein synthesis and is part of the amino acids - cystine and thiamine, which are responsible for the building blocks of proteins. Most hydroponics nutrient solutions come in a twin pack A and B; this is done to primarily separate sulphur from calcium. If mixed as a concentrate, the two nutrients conflict and create insoluble calcium sulphate gypsum.

Deficiency

Sulphur deficiency turns young leaves lime green to yellow. As the deficiency progresses, the leaves yellow interveinally and have a definitive lack of succulence. Veins remain green, the leaf stems and petioles turn a purplish colour and the leaf tips tend to burn then darken and turn downwards. The symptom of this deficiency is that with the older leaves it is more noticeable than the younger ones. This deficiency can also resemble the deficiency associated with nitrogen. If the deficiency is not addressed, it can cause elongated stems to become woody at the base of the plant. This deficiency is also enhanced if the pH is too high or when too much calcium is present.

Progression of sulphur deficiency is very similar to nitrogen deficiency; the older leaves of the plant turn pale green, the leaf stems change to a purple colour, more and more leaves change to pale green. After this, leaves change to pale yellow in colour.

Interveinal yellowing then occurs and if left, more leaves develop purple leaf stems and the entire leaf system yellows.

Treating a sulphur deficiency requires little effort. Simply change the nutrient solution for a fresh batch, pH-adjust it and away you go; within 2-3 days the deficiency will have been eliminated.

Toxicity

When sulphur toxicity occurs and the CF levels remain high, the plants will tend to absorb more available sulphur which in turn blocks uptake of other nutrients. Symptoms include overall smaller plant development and growth, it also encourages uniformly smaller, darker green foliage. The leaf tips and the margins of the leaf can discolour and burn if left unchecked.

Treating sulphur toxicity is once again easily rectified. Dump your nutrients reservoir, replace with pH-adjusted plain water with no nutrients, run this through your system for 2-4 days, then replace this reservoir with a fresh reservoir of nutrients at a slightly lower CF level than before, adjust your pH and away you go.

Zinc (Zn)

Zinc can be the most common micronutrient deficiency in dry and hot climates. Magnesium and zinc work together to promote enzyme functions. Zinc as well as other elements helps the formation of chlorophyll as well as preventing its depletion. High energy plants are more susceptible to zinc deficiencies.

Deficiency

Zinc deficiency is as stated before, the most common micronutrient deficiency. It starts with younger leaves showing signs of interveinal chlorosis. The newer leaves and the growing tips develop a type of small thin blade that contorts and wrinkles the leaf. Then, the leaf tips and later the margins of the leaf, discolour and tend to burn. It is very easy to get these symptoms confused with a lack of manganese or iron. However, when this deficiency is acute, the new leaf blades contort and dry out. The flowers also contort and develop in odd shapes; they can turn crispy and then dry and become hard. A definitive lack of zinc can stunt all new growth and prohibit buds and flowers.

Zinc deficiency progression starts with the young leaves developing interveinal chlorosis. Then, new leaves develop but the leaves are thin and weak. The tips of the leaves change to dark brown and then die back. Horizontal new growth then contorts, and leaf growth and new bud development stops.

Treating a zinc deficiency is very straightforward. Simply dump your existing nutrients and replace them with a new nutrients solution, balance the pH and after 2-3 days, the deficiency will have subsided and will have been eliminated.

Toxicity

Zinc toxicity, if acute, is extremely toxic. The plants in this situation will die very quickly. Too much zinc also acts as an iron prohibitor stopping the iron's ability to function, which in turn causes an iron deficiency.

Manganese (Mn)

This deficiency is common indoors. Manganese is engaged in the oxidation reduction related with the photosynthetic electron transport. Manganese acts as a catalyst for many enzymes and has a major role to play in the chloroplast membrane system development.

Deficiency

Manganese deficiency can be found first in the dying leaves; these leaves become yellow between the veins of the leaf. Then, interveinal chlorosis occurs with the veins remaining green. The symptoms then migrate to the older leaves once the younger leaves have been affected. As the symptoms progress, dead spots develop on the severely affected leaves before the leaves fall off entirely. The overall growth of the plant is stunted and the maturation and finishing of the plant may be prolonged. Acute deficiency can resemble a severe lack of magnesium.

Manganese deficiency progression starts with interveinal chlorosis on the young leaves. This then progresses to the older leaf systems. The dead spots develop on the very badly affected leaves resulting with the overall growth of the plant remaining stunted.

Treating a manganese deficiency, and you guessed it, is very easy indeed. Just dump the old nutrients, replace them with new nutrient solution, pH-adjust the solution and after 2-3 days, hey presto!

Toxicity

Manganese toxicity starts with the younger and new growth developing chlorotic dark orange to rusty brown mottled leaves. This is highlighted in the younger leaves before the older leaves start to develop the same symptoms. Overall vigour of the plant stops and the growth is lost. This toxicity can also be increased unknowingly by low humidity. Low humidity causes extra transpiration which in turn causes more manganese to be drawn into the foliage. The excess of manganese, again, also acts like a prohibitor and causes iron and zinc deficiencies.

Treating manganese toxicity is again very straightforward. Simply dump the old nutrients and replace with pH-adjusted water. Flush the system through for 1-3 days, then replace this solution with a fresh, lower CF value nutrients solution, pH-adjust and allow 2-3 days for the problem to be resolved.

Iron (Fe)

Iron is typically found in a chelated soluble form which makes it immediately available for consumption via the roots. Deficiencies indoors are not that frequent. Iron is crucial for the transportation of electrons during photosynthesis and respiration. It is also fundamental to the enzyme systems of the plant's growth. Iron acts as a catalyst for chlorophyll production. This element is crucial for nitrate and sulphur reduction and assimilation. Plants can sometimes have a difficult time absorbing iron, especially if the pH is high.

Deficiency

Iron deficiencies are only common when the pH of the reservoir is above 6.5 and very uncommon if the pH is maintained below this. Iron deficiencies may occur during rapid growth spurts or even stressed times. However, symptoms can disappear on their own.

First symptoms of iron deficiency appear on the smaller leaves as interveinal chlorosis, where the veins themselves remain green, but the area between the veins turns yellow. This interveinal chlorosis starts at the head of the leaf, not the tip, where the apex of the leaf is attached by the petiole. As the deficiency increases, the chlorosis becomes more severe. Leaves, in very bad cases, will drop off. Iron deficiency can sometimes be linked to an excess of copper, reducing the potential uptake of iron for the plants.

Progression of iron deficiency starts with the younger leaves and the growing shoots turning a pale green, which progresses to yellow in between the veins but starting at the head of the leaf. Veins still remain green. Then, more leaves yellow and develop interveinal chlorosis, spreading through the plant. Finally, the larger leaves yellow and also develop the interveinal chlorosis. If allowed to progress, leaves will develop necrosis and fall.

Treating an iron deficiency is once again very simple to implement. Get rid of the old reservoir of nutrients, mix a fresh batch increasing the CF slightly, pH-adjust the reservoir and away you go. Two to three days later all should be well. If this does not solve the problem, make sure that the water temperature is not too cold or chilly. The water temperature should be a few degrees less than the room temperature. If this is not the case, then invest in some water heaters and adjust the reservoir's temperature, just to eliminate the chill, not to heat the water. Warm water is more detrimental to the plant than cold water.

Toxicity

Iron toxicity is rare. High energy plants are not damaged by high levels of iron. However, high levels of iron can prohibit the plants' phosphorus uptake. So very high levels of iron can cause the leaves to turn bronze and encourages small, brown and dark leaf spots.

Treating iron toxicity is straightforward. Just dump the existing nutrients from the reservoir, replace with plain pH-adjusted water with no nutrients. Flush the system for one to two days, then remove this flushed water from the reservoir. Replace this with a fresh batch of nutrients with a CF value slightly lower than before, one or two points should suffice, then pH-adjust the system and the toxicity treatment is complete.

5
B
$[He]2s^2 2p^1$
Boron
10.81

Boron (B)

Boron, in hydroponics systems, usually causes no problems. These deficiencies rarely occur indoors. Scientists have collated some evidence that boron aids the formation of nule acid RNA uracil. More evidence also suggests that boron helps in the role of cell division, differentiation, maturation and respiration.

Deficiency

Boron deficiency starts with the stem tip and root tip mutating. Root tips discolour and root growth slows and elongation starts. Growing shoots also get a burnt look similar to excessive light burn when you get your lights too close to the plants. The top shoot mutates and burns, and is quickly followed by the lower growing shoots exhibiting the same symptoms. If left to progress, the tips die. Leaf margins discolour and also die back in spots. Between the leaf veins, necrotic spots develop. Roots often become soggy and the insides of the roots turn mushy which in turn becomes the perfect breeding ground for root rot and disease. The deficient leaves become mutated, thicken, wilt and chlorosis and necrosis set in.

Treating a boron deficiency is, and you guessed it, easy. Just replace the reservoir with fresh nutrients increasing the CF slightly by one or two points, pH-adjust and the job is done.

Toxicity

Boron toxicity manifests itself within the leaf tips by yellowing, then as the toxicity progresses, the leaf margins become necrotic in the centre of the leaf. If left unchecked, the leaves will fall off.

Treating boron toxicity is not complicated. Just dispose of the old nutrients from the reservoir.

Replace with plain pH-adjusted water, flush the system for one to two days then replace the reservoir with a fresh batch of good quality hydroponics nutrients but at a slightly lower CF value than before (one or two points), pH-adjust the solution and away you go.

Chlorine (Chloride) (Cl)

Chlorine deficiency is very uncommon in a hydroponics garden. Chloride is found in most water supplies. High energy plants can tolerate low levels of chlorine. This is normally not a component of nutrient solutions, as chloride is not in short supply in the indoor garden due to sufficient levels in tap water, but it is fundamental to the photosynthesis and cell development in roots and foliage. Chloride also increases osmotic pressure in cells, which in turn opens and closes stomata to regulate the moisture flow of the plant.

Deficiency

Chloride deficiency is not very common. However, if this occurs, young leaves turn pale and wilt. As the deficiency increases, leaves become chlorotic; they also become bronze in colour. Roots mutate and develop thick tips and end up stunted.

Treating chloride deficiencies is not difficult; simply add chlorinated water to the nutrient solution.

Toxicity

Chlorine toxicity starts with the younger leaves developing burnt tips and margins and the leaf systems will turn a bronze colour. This is most common in very young seedlings and clones, as these are most susceptible to this problem. If left unchecked, the burnt tips will migrate through the entire plant and yellowish, bronzed leaves will dominate, which in turn will be smaller and grow at a prohibited rate.

Treating chloride toxicity does not involve much work; flush the system through with non-chlorinated water, or allow the reservoir of water to stand overnight giving it an occasional stir. This will allow the chlorine to evaporate into the atmosphere and reduce the levels of chlorine in your water.

29
Cu
[Ar]4s^{1}3d^{10}
Copper
63.55

Copper (Cu)

Copper, apart from its fungicidal properties, is a component of a number of enzymes and proteins. Absolutely necessary in very minute amounts, copper helps in carbohydrate metabolism and nitrogen fixation.

Deficiency

Copper deficiency again is very unusual in high energy plants in a hydroponics garden. However, if this does occur, then the young leaves and growing tips are the first to be affected and end up wilted and limp. The leaf tips and margins are next to show signs and develop necrosis and turn a dark, coppery grey. Sometimes, if this is left unchecked, an entire copper deficient plant will completely wilt, even if watered adequately. The growth is slow if the plant has copper deficiency, with decreased yield.

Treating a copper deficiency is a straightforward process. Dump the existing nutrients and replace with a fresh batch of nutrients raising the CF level slightly, pH-adjust the solution and allow 2-3 days for the recovery.

Toxicity

Copper toxicity is more prevalent in a hydroponics garden. Although copper is essential to high energy plants, it is also extremely toxic to these plants if subjected to high levels of it. Firstly, the toxicity slows the overall progress of the plants' growth. As the toxicity increases, symptoms include interveinal iron chlorosis deficiency, stunted overall growth, sick roots darkening off and slow to grow with a mutated thickness.

Treating a copper toxicity is simple. Dispose of the existing nutrients and replace with pH-adjusted plain water. Flush the system through for 1-2 days, then replace this solution with a fresh batch of nutrients but at a slightly lower CF value than before, pH-adjust, then allow 2-3 days for recovery.

Molybdenum (Mo)

Molybdenum deficiency is rarely found in a hydroponics garden. Molybdenum is one of two major enzymes that is responsible for converting nitrate to ammonium. High energy plants find this essential but use this in very small quantities.

Deficiency

Molybdenum deficiency is typically never found in high energy plants. However, if this does occur, then firstly the older leaves and middle-aged leaves yellow and some leaves will develop interveinal chlorosis. The leaves continue to yellow and then develop rolled up margins as the deficiency accelerates. If allowed to progress, extreme symptoms are mutated twisted leaves that die and fall off. The overall growth of the plant is stunted.

Treating a molybdenum deficiency is the same as treating the other deficiencies. Just replace the nutrients with a fresh batch but increase the CF slightly, pH-adjust and the job is done.

Toxicity

Molybdenum toxicity is again seldom found when growing high energy plants. However, if this does occur, the toxicity will cause a deficiency of copper and iron as the concentrations of molybdenum will restrict the plants' uptake of these nutrients.

Treating molybdenum toxicity is easy. Get rid of the existing nutrients, then flush the system through with plain pH-adjusted water for 1-2 days, replace this solution with a fresh batch of nutrients but at a slightly reduced CF level, then pH-adjust and away you go.

Silicon (Si)

Silicon is already available in most water supplies. In high energy plants this does not cause any complications due to excessive deficiencies. Silicon absorbed as silica by the plants is found in the epidermal cell walls and it is stored in the form of hydrated amorphous silica. It is also stored in the walls of other cells. A definitive lack of silicon has been proved to decrease overall yields and vigour of some fruit or flower bearing plants. It can also cause leaves to mutate. This product has been shown to strengthen and reinforce the cell structure of the plant. This can benefit the plant in many ways, including making it more difficult for plant feeding pests to take hold as the double hard cell structure of the plant prohibits the pests from biting into the soft cell structure of the leaves or stems. Silicon is still being studied and high energy plants are the focus in this research.

Different Hydroponics Nutrients

Single Pack Nutrients

These are relatively new to the hydroponics market and are very popular, however, they are popular due to their ease of use over their ability to perform. They are basically a two or three part nutrient watered down so much that apparently, the differing nutrient concentrations do not conflict with each other. However, most single pack nutrients recommend using a bloom boost in the flower cycle of the plant's life. This has the missing conflicting nutrients that the bloom nutrients on their own do not have. So in effect, during the bloom cycle, you are fundamentally using a twin pack nutrient although it is marketed as a single pack nutrient. Personally, and in general, these single pack nutrients do not perform as well as a good twin or triple pack. The reason for this is that even if the single pack is diluted enough, you will still get some conflict on a macro and micro level. Even if this conflict occurs on a very small scale, this will prohibit the potential of the plant's ability to perform. For years, top nutrient manufacturers have been making twin and triple pack nutrients and the reason for separating the nutrients is purposeful i.e. they would have not made twin or triple packs unless it was necessary to do so, due to the nutrients conflicting. With this in mind, it seems early days for the world of single pack nutrients, so if you are thinking it's easier this way therefore worth doing, think again. At the end of the day, when in flower, you will be in effect using a twin pack nutrient, so if this is the case then do so from the start. In general, single pack nutrients are easy to use, do not go a long way and give mediocre results.

Example of Single Pack Nutrients

Twin Pack Nutrients

These nutrients are by far the most readily available and most commonly used nutrients on the market. They are called twin pack but are actually quad pack nutrients as they have a separated twin for the grow stage and a separated twin for the bloom stages, i.e. they consist of a Grow A and B and a Bloom A and B. So, in total, you get 2 bottles for the vegetative cycle and 2 bottles for the bloom cycle. These nutrients are very simple to use and nobody should be put off due to any misconceived complexities. In a nutshell, you use equal amounts of the A bottle as of the B bottle. So, during the grow stage, use the same amount of Grow A and the same amount of Grow B. Likewise, during the bloom cycle, stop using the Grow nutrient and swap to the Bloom nutrient and use in equal amounts. Really, if this is the reason that people buy single packs instead of twin packs, then they should not be bothering with hydroponics in the first place. The reason for dividing the nutrients into a twin for the grow cycle and a twin for the bloom cycle, is to ensure that the right nutrients are available during the vegetative cycle of the plant's growth and that the right nutrients are available during the flowering cycle of the plant's life, as the plant requires different levels of N-P-K during the different cycles. The other reason for splitting them into four is to ensure no nutrients lock or conflict, which will or can occur. This is the tried and tested model for nutrient manufacturing and the most widely used nutrient available with also the most choice in availability. Most hydroponics retailers stock at least 4-8 different makes of twin pack nutrients. These nutrients are also made at a higher level of concentration than the single packs which means that the bottle will go at least twice as far, if not more. So, you will not need to keep going back to the hydro shop to buy more. It is still advised that you employ the use of a bloom boost nutrient typically P-K13-14 during the later stages of flowering with these products, as during this time frame high energy plants need extra levels of this supplement and as they only require it approximately three weeks before harvest, it therefore cannot be pre-mixed into the bloom A and B formulae.

Personally, the best results I have ever witnessed came from twin pack nutrients. They are very easy to use, go a long way, and give great results.

Example of Twin Pack Nutrients

Triple Pack Nutrients

Triple packs are not so readily available as the twin packs and only a handful of manufacturers make them in this fashion. However, those that use the triple pack normally are sold on it and stick to it like glue. These are, as they are aptly named, a triple pack of nutrients. They consist of three parts Grow, Bloom and Micro and again, the reason for separating these nutrients into three parts is to stop any possible conflict or nutrient lock. The way this nutrient works is that during the vegetative cycle of the plants' life, you use a combination of the grow, bloom and micro parts but use more of the grow and less of bloom. During the flowering cycle, you again use differing amounts of three nutrient bottles but this time using more of the bloom and less of the micro. In fact, during the different weeks of the plants' development, you can use different levels of the three nutrient bottles, giving them exactly what they need, when they need it. As you have probably worked out, this is the most complicated way of feeding your plants properly, however, once you have mastered it, it is actually very simple indeed, but with an air of complexity. So, you can really fine-tune it if you so wish. These nutrients, like the twin packs, are very concentrated and therefore again go a long way. They also deliver great results when properly administered and allow the advanced grower to fine-tune the nutrients for a prized crop. The very advanced grower even uses the triple pack as complementary nutrient for the twin pack i.e. the grower primarily uses the twin pack nutrients but adds a little of the triple pack to the mix to achieve the results they need. In general, triple pack nutrients are a little tricky to use at first, go a long way and give great results.

Example of Triple Pack Nutrients

Hydroponics
Indoor Horticulture

Chapter 6

Oxygen and Air

Ventilation of air and the supply of fresh air is sadly one of the most overlooked factors when creating a grow room. Air circulation and ventilation are absolutely definitive factors for happy plants and great yields. Without the availability of fresh air, the stomata become restricted and so does the plants' potential.

As with humans, plants need oxygen to survive. In fact, cell for cell compared to humans, plants use very similar amounts of oxygen. So, in effect, plants need and use lots of oxygen. It has been found that a dried plant will consist of approximately 45% of oxygen atoms. So in conditions where the air has become stagnated inside a grow room and where the air has less than 20% oxygen, plants struggle to perform.

Plants, obviously as a by-product of photosynthesis, produce oxygen. During the light cycle, plants are in effect breathing out oxygen so the leaves have relatively easy access to oxygen during the daylight cycle. The roots on the other hand have a much harder time to utilise enough oxygen for their requirements. This results in restricting root respiration, which in turn slows photosynthesis, which in turn reduces the growth potential of the plant.

Consequently, the happiness and the potential of the plant are very dependent upon the roots securing enough oxygen. Plants are only able to grow as well as the roots allow it; yield can be directly measured in proportion to root growth.

Oxygen and Water

These are the facts; warm water holds less oxygen than cold water, however, very cold water can be detrimental to the plants, effectively shocking the root system until the water warms up enough for

Returning nutrient solution to the reservoir can provide good aeration

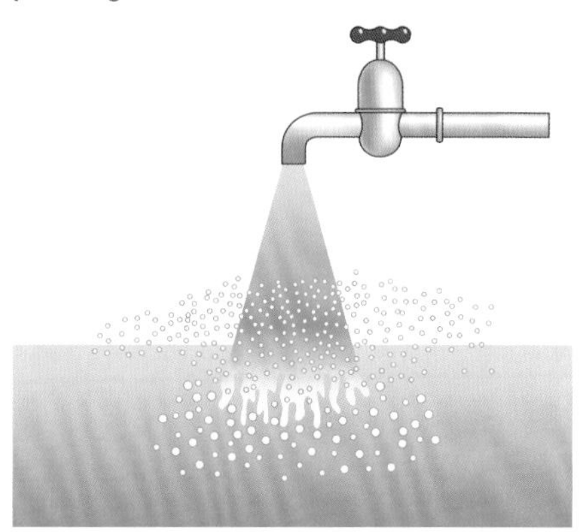

Tap water being administered to the reservoir provides good nutrient solution aeration

the plants to get over this shock. On the other hand, very warm water has a detrimental effect on a plant, again shocking the plants' roots until it cools to a level that enables the roots to recover from the shock. So, the colder and fresher the water, the more oxygen content; the warmer, the less.

Approximately 0.0014% of dissolved oxygen can be found in very cold, fresh water. Approximately 0.0008% of dissolved oxygen can be found in fresh water at room temperature and water at room temperature is exactly how the plants like it. Approximately 0.0005% of dissolved oxygen can be found in fresh water at 86°F or 30°C. So, as illustrated below, heating your water too much can be very detrimental to your plants' potential.

Oxygen obtained via the roots directly from the nutrient solution only makes up 1% of the plants' needs. So the oxygen supplied via the nutrient solution is only a minor source supplied to the roots. Aerating the nutrient solution will help, however, this mainly serves to kill off any pathogens, basically keeping the nutrient fresher for longer. So you can see that allowing the plants some dry time will serve to get more air to the roots. Even in aeroponics, which delivers a highly oxygenated solution directly to the rootball, it is essential to give plants dry time (at night for example), to allow air to reach the root system. Another technique is to add H_2O_2 to the reservoir, which again, due to its extra oxygen atom, increases the dissolved oxygen in the reservoir. However, when H_2O_2 is diluted, it becomes unstable and soon breaks down and dissolves completely into the reservoir, so it normally will only oxygenate the reservoir for a few hours. Also, when H_2O_2 is regularly administered to the reservoir, it keeps the nutrient solution fresher for longer. However, overdoing H_2O_2 can be very detrimental to the food chains in your reservoir and detrimental to the plant's roots. In high doses, it attacks many usable food chains that the plant will be deprived of.

Highly oxygenated root systems benefit from another very important fact. Oxygen manipulates and affects the electrical charges found in water and for that fact, nutrients. This change in electrical charges allows roots to uptake water and

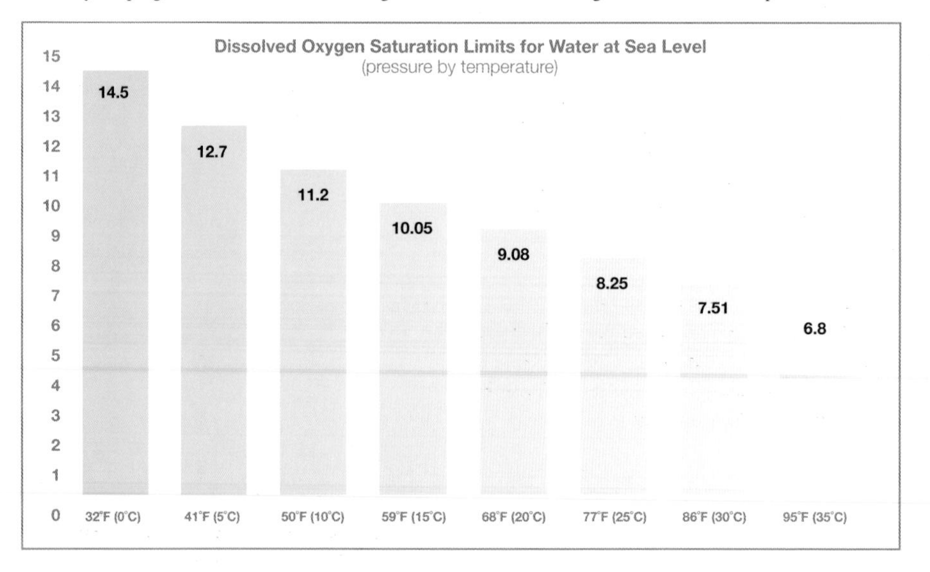

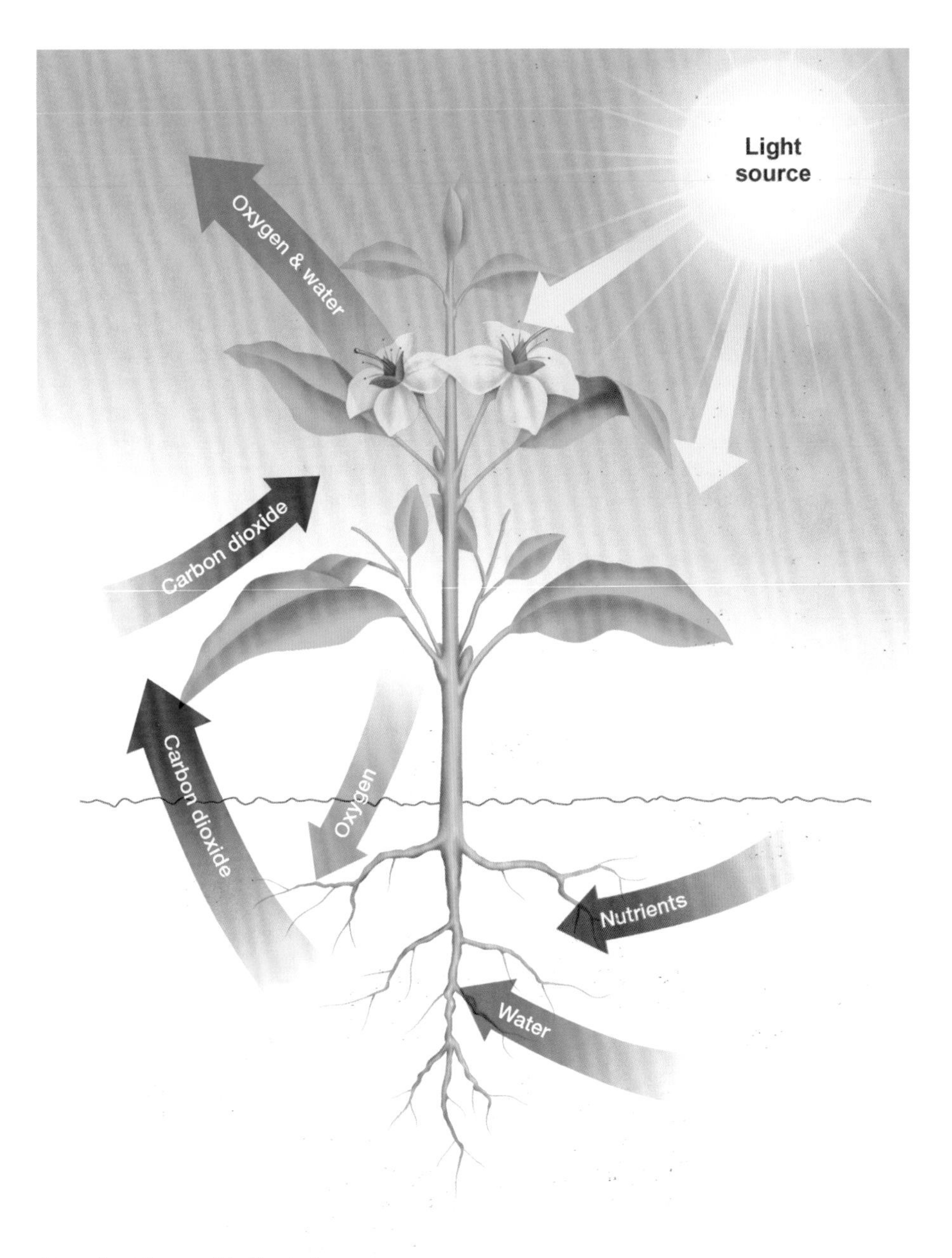

Example of Autotrophic Metabolism

nutrients with a lot less energy compared to non-oxygenated root systems, so clearly roots benefit from as much oxygen as you can provide them.

Properly aerated nutrient solutions will absorb up to 0.0008% oxygen and hold it for up to 24 hours.

Oxygen

Like humans, plants actually need to respire 24 hours a day to survive. During the light cycle, plants will uptake faster if there is an abundance of oxygen available. As stated before, during the light cycle, the plants are constantly producing oxygen so this is not normally a problem. However, the roots of a plant, day or night, do not produce any oxygen. The plants' internal oxygen is manufactured from the leaves splitting hydrogen from the water molecule to generate sugar, which then releases excess oxygen. When the plants are in the night cycle, then they are dependent on the air around them to obtain their oxygen.

During this night cycle, plants' respiration drops to a steady ambient level of maintenance. This is due to the plants not wanting to use any of their stored sugars during the night cycle any faster than they are required to do so. Consequently, during this cycle, surrounding oxygen is crucial as the leaves are not processing any. However, during the night cycle, the speed of respiration is not actually controlled by oxygen levels but by the temperature of the grow room; nevertheless, it is still very important to allow fresh air to the plants during the dark period. The rate at which plants are respiring is also the rate at which the plants are growing; these two factors are completely temperature dependent.

Air Exchange

The use of fresh air is central to the productiveness

Examples of Extraction Fans

of any grow room. The concept of a grow room is, in simple terms, to replicate as well as possible, and in some cases, to improve upon the conditions outside that a plant is subjected to. A grower is therefore mimicking and even super charging what one would naturally find outside, inside; this is the key to very productive plants. As you can imagine, air exchange inside a grow room is a key and fundamental part of a successful grow room. Not only is it essential for the supply of fresh oxygen for the plants, but also for the replacement of fresh CO_2 that is found naturally in air. Carbon dioxide and oxygen are simply basic building blocks for plant life. Plants uptake oxygen and CO_2 to manufacture sugars which are used to fuel growth. If you get limited oxygen and CO_2 to the grow room then the plant's growth will slow to a crawl.

This situation is very easily rectified and is also inexpensive to buy and employ. Investing in a good extraction fan is all that is required. An extraction fan should be mounted at the highest point of the grow room so as old, hot air rises, the extraction fan will suck this old, hot humid air out and through this action, creates a positive vacuum in the grow room. Making some inlet hole for fresh air to come in at a low level will make sure that air exchange can be maintained for your plants.

Employing an extraction fan has many benefits. It ensures a good supply of fresh air and therefore, a good supply of oxygen and of CO_2. It also removes humidity, stale, stagnated air, and heat. If you do not employ extraction, then you can liken the grow room to an oven with the light becoming the heat source within the oven. If you do not vent out hot air, the hot air is allowed to build up, just like in an oven, and slowly but surely, the grow room temperature will rise to levels that become out of control, and if left, can result in no growth for your plants or even complete annihilation of your living garden. So, as hot air rises in the grow room, it is

Examples of Fan Controllers

absolutely essential to vent this hot air out. Through venting the hot air out and due to the vacuum effect, fresh air coming in from lower vents will reduce the overall temperature inside the grow room. In this way, you can maintain a level that best suits your plants.

This temperature control is also essentially linked to humidity. Again, if you do not employ an extraction fan, as with the oven effect described above, humidity will also follow suit and rise. The end result is high humidity, which in turn like too much heat, can cause your plants many problems and difficulties. So typically, if you get temperature under control, humidity will also fall into line. Both factors are critical for getting the most out of your plants.

Plants usually like temperatures in the grow room to be around 75°F with a humidity of around 50% which with the use of extractors can easily be achieved. It is also important to note that day and night time temperatures should be similar. It is recommended that you have no more than a 10° swing from day and night temperatures. So, if you are 75°F during the day, you should not be less than 65°F during the night. This ensures that the plants are not subjected to temperature stress or shock. If the grow room temperature plummets during the night cycle, the shock sustained to the plants is massive, often resulting in plants not doing any growing until they have overcome the shock, way after the temperature has risen to the desired level. That is to say, the plants can take 4-6 hours to get out of shock even though the temperature has been right for this period of time, with the end result being that the plants' whole cycle can be slowed down for months, if left unchecked. As you can imagine, a plant that is not growing for 6 hours during an 18 hour day in order to recover from shock and stress, slows its whole growth rate down by approximately one third. So, an 8 week maturing plant will in fact take up to 12 weeks to mature, and that is if the stress does not do anything else strange to the plant. In short, keep the day and night time temperatures as close together as you can.

Exact temperature control can be achieved by utilising thermostats on your extraction fans and on your heaters. These are excellent aids to an indoor grow room. Thermostats can be used on your fans to turn them on when the temperature gets too high, then once the fan has lowered the temperature to the desired level, the fan turns off again waiting to turn back on when the temperature gets too high once more. In contrast and additionally, you can get thermostats that control heat which turn the heaters on when the temperature gets too low, lets the temperature reach the desired level and then switches the heaters off, waiting to turn back on when the temperature once again drops. If you use both of these thermostats, you can maintain a grow room quite easily with only a 5° difference in temperature.

It is not widely known that an extraction fan sucking hot air out is at least four times more efficient at complete air exchange within your grow room than an extraction fan used to push air in. When employing an extraction fan, it should be able to replace all the air in the grow room, that is a complete air exchange, in less than 5 minutes. For this reason, you need to know the length, the width and the height of the grow room to work out what size extraction fan to go for. All extraction fans are sold on their square cubic metre capacity, which makes it very easy to get the right fan for the job.

It is also worth noting that the more light you employ, the more heat you will generate, and the more extraction you will need to use, so with this in mind, always over calculate the size of fan you require. That said, the bigger the fan you use, the louder the fan and air noise is. If noise is a concern, and in many indoor grow rooms noise is always an issue, then you can fit very effective silencers directly to the fans which act as a baffle and really do reduce the noise. If this is not enough, you can fit carbon filters to the extraction fans, which most growers do, as the active carbon removes any unwanted odours before it is expelled outdoors. These carbon filters reduce noise and remove all odours and when coupled with a silencer, completely shut up even the noisiest fans. In conclusion, fresh air is critical to healthy plants for replacing the oxygen and replacing the carbon dioxide. Maintenance of temperature and humidity again, are critical for healthy plants.

Examples of Fan Filter Combinations

Example of a Silencer

The simple and inexpensive solution of utilising an extraction fan is an absolute must for an indoor grow room. If you do not use one of these, then don't be surprised if it all goes pear-shaped. Do not think twice – you do and will need them. Combine them with heaters and thermostats and you have got complete day and night time temperature and humidity controlled, so you can keep your grow room just how your loved plants like it.

☺

Hydroponics
Indoor Horticulture

Chapter 7

Carbon Dioxide

It is not widely known that the dry matter in a plant is 90% carbon, hydrogen and oxygen. The majority content which is carbon, is taken into the plant through carbon dioxide available from the air.

The average air that we inhale contains 0.003-0.004 % of carbon dioxide. Carbon dioxide is an odourless, non-flammable, colourless gas.

It is believed that the prehistoric plants that developed aeons ago had environmental conditions with very high levels of carbon dioxide in it. In today's plants, due to their evolution, these more modern incarnations have maintained their capacity to harness more CO_2 than the current environment now has.

Science has shown that high energy plants have the ability to consume PPM levels of 1200-1500 and in some cases even as high as 1750 parts per million; the average air we breath contains 300 parts per million. So, as you can see, high energy plants in

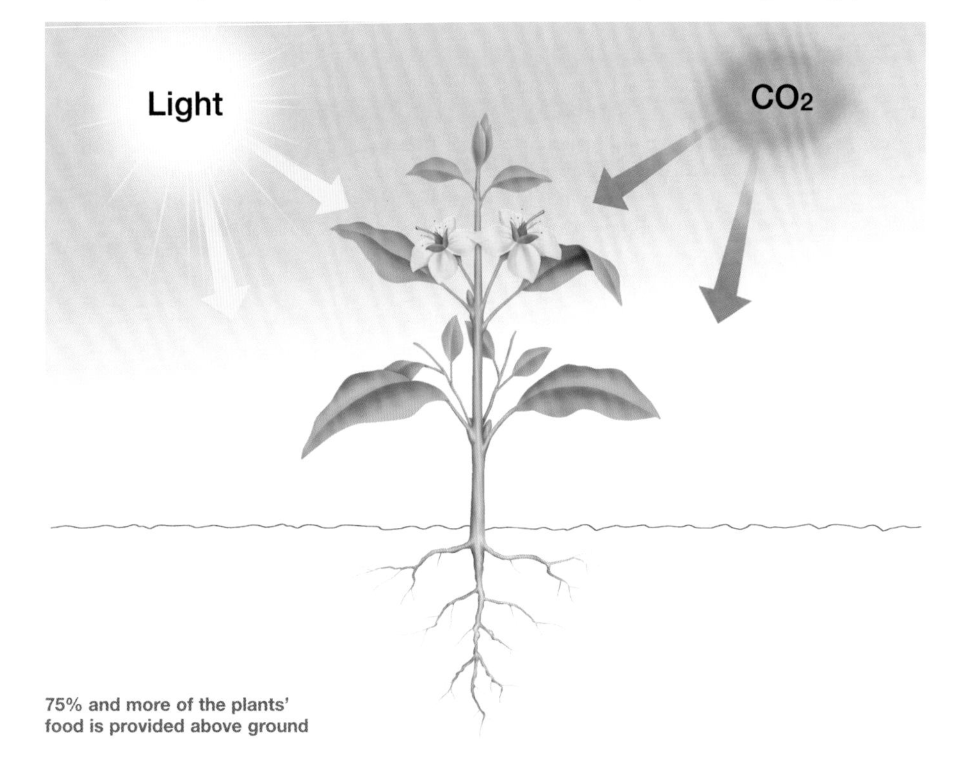

75% and more of the plants' food is provided above ground

the right conditions can consume much more than nature is currently providing.

The consumption levels of CO_2 that the plants need is directly correlated to lumens that they are subjected to i.e. the greater the intensity of light, the greater levels of CO_2 that the plants can consume, the greater the plants become.

Again it is not widely known that a plant saturated in light will only grow to be as big as the CO_2 levels in that particular environment. In other words, you might have all the light a plant could ever want, however, if the plant is only receiving a little CO_2 then most of the light available to the plant is simply going to waste and is not available for consumption via the plant, as it cannot utilise the CO_2 to photosynthesise the light.

An average high energy plant in an average lit grow room with no in or output fans will consume the available CO_2 within that grow room in a few hours. So any growth after all the CO_2 has been utilised will slow completely down to a snail's pace.

This uptake of CO_2 is only necessary during the lighting cycle; plants do not require CO_2 during the night cycle when the lights are off. Plants, like humans, do breathe out CO_2, however, during their daylight cycle, the plants utilise the CO_2, manufacturing sugars to feed the plants and so during the day cycle appear to only breathe out oxygen.

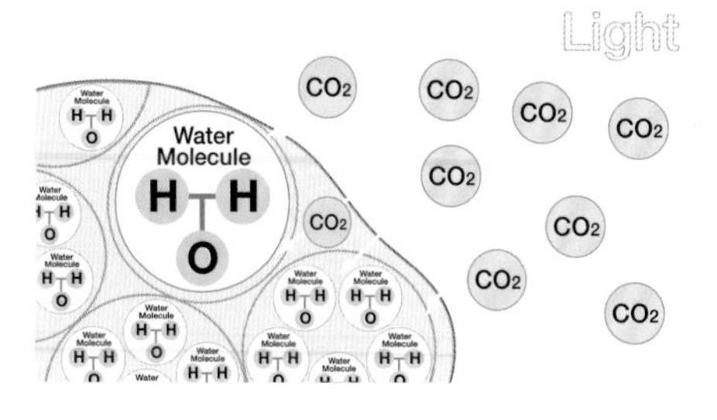

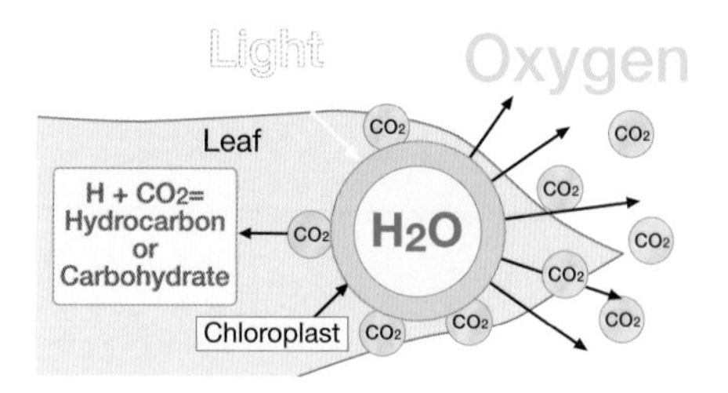

Science has shown that the more light a high energy plant is subjected to, the more CO_2 the plant requires for photosynthesis. During this photosynthesis, it takes approximately 10 photons to generate enough electrons to create sufficient energy to split one carbon dioxide molecule into carbon and oxygen atoms to form sugar. Typically, trillions of these photons are hitting a plant's leaves, which means if the grower is not providing enough CO_2, then most photons will simply bounce off the leaves without being consumed.

CO_2-Enrichment

Carbon dioxide enrichment has been in wide use in commercial greenhouses for more than 25 years. Delivering more CO_2 to a grow room with high energy plants in it, will stimulate growth. High energy plants in good lighting conditions consume 1200 to 1500 parts per million. However, science has shown that if you overdo it with CO_2 then it's like overdoing anything in hydroponics and can have a detrimental effect. If you get all the environmental conditions right and are injecting CO_2 at the right levels you can get to a situation where you are doubling the plant's growth. So, as you can imagine, this is why CO_2 is so widely used in an indoor grow room.

High energy plants in full sunlight, which equates to approximately 5000 lumens per square foot, can process 1500-2000 parts per million of CO_2. However, indoor gardens with the light levels of approximately 3000 lumens per square foot need approximately 1000-1500 parts per million of CO_2, compared to normal air which is typically 300 parts per million, so you can see that CO_2-enrichment is worthy of investment and use.

Carbon dioxide enriched high energy plants, due to their enhanced growth patterns, require higher maintenance than non CO_2-enriched plants. CO_2-enriched plants use more water, nutrients, and for that matter, more space. They can consume the aforementioned up to twice as fast as non CO_2-enriched plants. The temperature can also be slightly increased in a CO_2-enriched grow room.

People, for some reason, get somewhat confused about CO_2, believing that the roots need to uptake it. This is simply not the case. It is true that the roots benefit from the leaves uptaking CO_2, however, injecting CO_2 to the root system will have nothing but detrimental effects. Roots require oxygen; CO_2 will suffocate them.

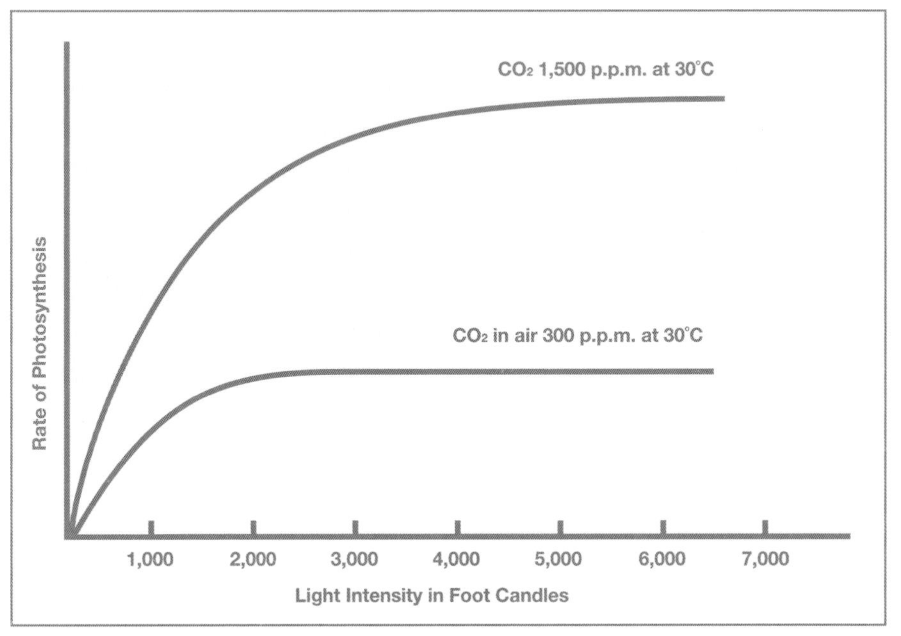

Increased Rate of Photosynthesis with Increased CO_2 Concentration and Light Intensity

CO_2 Release Systems

There are two main methods of generating CO_2. Burning propane gas or injecting compressed bottled CO_2 via an emitter system. They both have their pros and their cons.

Propane Gas Burner

Let's start off with the propane gas burner. These are designed primarily for generating carbon dioxide formed as a by-product of burning the propane gas. These little beauties can generate a lot or a little usable CO_2 for your plants. They are very similar in design to a gas burner with a pilot light which is connected to a solenoid, and via a timer the solenoid is switched on and the propane gas burner ignites fully thereby burning propane to generate a lot of CO_2. These burners have been specifically tweaked to produce a maximum amount of CO_2 from the burning gases and although they have been modified, they obviously produce another by-product, which in rare cases is a blessing and in most cases a curse, and that is heat. Another blessing or curse with this type of CO_2-enrichment system is that another by-product caused by burning propane is water. Water created by burning propane will automatically raise the levels of humidity in the grow room. In short, you will have raised the temperature, and raised the humidity in your grow room through using a propane gas burner. This can, as stated earlier, be a benefit to some growers and a hindrance to others, depending on the environment of the grow room. However, this method allows you to generate a massive amount of CO_2. The reasons that these devices are so popular with indoor gardeners is that they are relatively inexpensive to buy but more importantly very inexpensive to run. The reason for this is that propane gas has been widely used in a domestic and commercial capacity for years for many applications apart from horticulture. So the availability of the propane gas bottle is huge and the cost minimal. In some grow rooms, these little beauties really do help crank it out. However, in other grow rooms, due to temperatures rising and humidity increasing, they are not suitable. One more drawback that is easy to overcome is that due to the fact they have a pilot light, they can generate too much light during the night cycle of your plant's life. This can confuse the plant and stop it from inducing fruit or flowers during a 12 hour cycle. Most growers, when using a propane gas burner, get around this by building a partial box around the burner to obscure any light output the pilot light might generate. One other drawback of excessive use is the by-product of nitrous oxide that can get to toxic levels if you overdo it with these machines. However, this can be avoided through proper use. These burners typically come with a table of room dimensions to burning times required to generate 1200-1400 parts per million. In order to get the correct CO_2 levels, you would simply apply this calculation.

Example of a CO_2 Burner

CO_2 Bottled Gas Release System

The other option for generating controlled carbon dioxide injection is an emitter system, which works from compressed bottles of pure CO_2. These systems are almost risk free and produce no toxic gases, no heat and no water. But you guessed it; they are neither cheap to buy nor cheap to run. These units also allow very precise, metered delivery of exact amounts of CO_2 directly where you want it. They are available as two types – a manual continuous flow option or an automatic short bursts option. Bottles of CO_2 gas can be obtained from all good hydroponics stores, or direct from an industrial manufacturer of these bottles and gas. You normally have to pay a deposit and rental on the bottle, then after this payment you simply pay the refill costs. Typically, the bottle you have purchased is swapped in for a full bottle every time you need it to be refilled.

The manual flow system works by way of a solenoid valve controlled with a short range timer that regulates the flow of CO_2 into the grow room. The solenoid valve is operated electronically via the timer and is used to start and stop the flow of gas from the regulator. This short range timer opens the solenoid valve for brief periods of times in the light cycle. Normal timers are not suited for this application and their on and off times are generally too long, injecting far more CO_2 than the plants require. The manual methods mean you have to work out the flow rate and the injection duration of CO_2 to obtain optimum conditions. You have to work this out with a complicated formula. Divide the number of cubic feet required by the flow rate. The cubic feet measurement is obtained by measuring the length, width and breadth of the grow room and times them together. For example, if the flow rate is set at 10 cubic feet per hour then the valve will need to open for 0.1 hours which is 1 divided by 10, or in English – 6 minutes, as this equates to 0.1 hour x 60 minutes. This will bring the room up to 1500 PPM.

As CO_2 is heavier than air it is likely to leak out from the grow room. On average, the CO_2 levels will deplete to 300 parts per million in approximately 3 hours due to leakage and plant's usage.

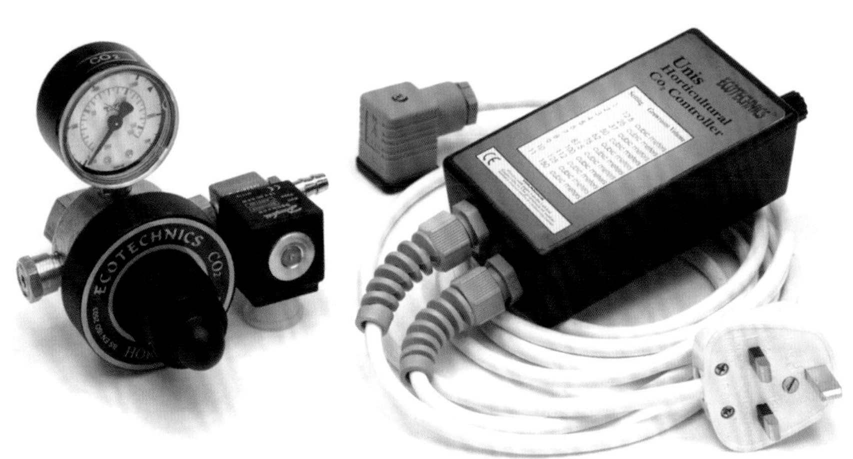

Example of a CO_2 Emitter System

When calculating the room size, do not calculate the entire room size because as stated before; CO_2 is heavy therefore falls to the ground then fills the room up like a swimming pool. Therefore, you only need to measure the room size at the finishing height of the plants which is typically half the actual room size. It is also advised to split the amount of CO_2 released, into smaller increments i.e. instead of one big burst, divide this into four so you have four short bursts per hour and this will maintain a more steady level of CO_2 in your grow room. If the pressure changes on your bottle, you have to recalculate and change the times to compensate for the loss of pressure.

The other option of an automated CO_2 release system works by having an already calculated injection rate by means of a controller that is linked to a regulator via a solenoid valve. With this system, you simply work out the square volume of the area you want to inject, dial this measurement into the controller and this gadget does the rest, preset at a level of 12000 parts per million. The very latest one even comes with a built-in light cell so that the CO_2 regulator even switches itself off during the night cycle. These systems, although obviously more costly, simply take the grief out of CO_2 injection. They do all the calculations for you and once set, will inject CO_2 almost every other minute in very short and measured bursts directly to your high energy plants.

The way the CO_2 is delivered to the plants is typically done via a small perforated delivery tube. You would suspend the tube, making sure it is looped in a ring from your CO_2 regulator, to ensure even delivery through the whole pipe system above your plants. This tubing carries CO_2 from the supply tank to the centre of your grow room above your plants. It has to be above your plants because, as mentioned previously, CO_2 is heavier and cooler than air, so it actually acts more like water than a gas. As soon as it is released, it falls on top of the plants. This method of releasing compressed CO_2 gas is virtually risk free and very precise, however, is expensive to buy and expensive to run, so larger growers tend to steer clear of it. However, when you compare the increase in yield, CO_2 is quickly becoming a must for a supercharged garden.

Example of a CO_2 Bottle

Chapter 8

Water pH

Water pH is a measure of the acidity or alkalinity of your water, be it in your system or out of your tap. This measure allows you to confirm whether the solution is too acidic for your plants, or too alkalinic; the measure reads from 0 to 14. Zero is the most acidic, 14 being its opposite and the most alkalinic, with 7 being neutral. The majority of nutrients used by a plant are soluble only within a limited range of acidity. This range is between 6 and 7 if you are growing in dirt, or 5.5 to 6.5 in most hydroponics systems to allow sufficient compensation for the particular grow medium used i.e. rockwool has a slightly higher alkalinity than clay pebbles, so when irrigating rockwool, you compensate for this by giving the medium a solution of around 5.5 on your pH level. In pebbles, it will be 6 to 6.5 for the same reason.

Ok, should your reservoir become too acidic then the dissolved nutrients in the water will precipitate and you will get nutrient lock, therefore, the necessary nourishment will be unavailable to the plants. This is also the same if the solution gets too alkalinic. Tests have shown that a plant subjected to conditions where the pH is too low typically end up very small and only growing, if at all, at a snail's pace. The same tests have shown that plants subjected to high pH are very pale with also the same slow, stunted growth characteristics.

pH can be measured using simple colour changing chemicals such as litmus paper or more accurately measured using a digital pH meter available from good hydroponics or aquarium retailers. All water has a pH value which can be measured using the above.

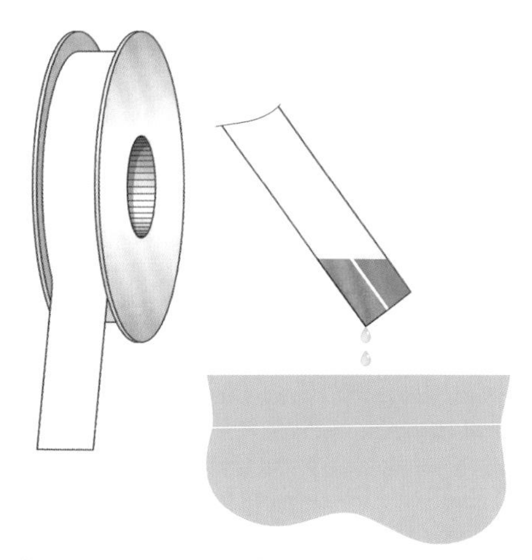

Example of Litmus pH Paper

After adding nutrients to your reservoir, you should remember that the nutrient employed will affect and change the level of pH in the reservoir, so always stir the reservoir thoroughly after adding nutrients and then allow it to rest before testing the pH and adjusting the level in your reservoir. If you find the level of pH is too high in your system then you should adjust it with some pH down which typically consists of phosphoric acid at 81% and is again available at your local hydro store. You should only employ a tiny drop at a time, then stir and re-test before adding any more as this stuff is extremely concentrated and takes a little time to mix properly. If, on the other hand, you find that the solution is too acidic, then you need to employ some pH up – this is usually potassium hydroxide at 50% and again available at your local hydroponics store. Like the pH down, use very sparingly and mix it well before re-testing.

Never mix these two solutions directly as these two chemicals are diametrically opposed and when mixed together as a concentrate, they will conflict and possibly explode.

It is also worth noting that if you put too much pH down in, it is not good practice to readjust it with pH up as both of these chemicals, when heavily employed, knock out very important food chains, so the more you use, the worse it will be for your plants as more good nutrients in the reservoir are destroyed. So if you get it wrong, simply throw the stock solution away and start again, being more careful and don't overdo it this time.

On this note it is once more worth mentioning that if you are running a re-circulating hydroponics system, like most growers in the UK, then you should be aware that in hard water areas where you have to employ much more pH down to get the pH of the water to the right levels, that it is not advisable to dump nutrients on a regular basis, as every time you do this, you not only need to fill the reservoir with more fresh water and nutrients but you also have to use a lot of pH down on every occasion. In so doing, you are knocking out and destroying food chains that the plants want as you have to use a lot of pH down to get the solution to the correct level. On the contrary, when you simply top up an existing reservoir with fresh water, do a little bit of pH balancing and the food chain is hardly affected. So, by replacing the solution in the reservoirs, you think you are giving your plants fresh and therefore more food when in reality, because you need to add so much pH down as the solution is not already buffered, you are in fact destroying much of the usable food chains. However, if you were simply to top up and replace the used water and nutrients, then readjust the pH, you will actually be helping your plants out more as not many food chains are lost in this process.

It is advised on new systems to check your pH daily, and once the system has been broken in and buffered, then every 2-4 days should suffice. Plant growth fluctuates the pH in a reservoir as they deplete differing nutrients from the reservoir, so bear this in mind. Also, tap water over periods of time also fluctuates in pH levels, so do not always think that your tap water is a particular level of pH, as some days it is not.

If possible, it is also advised to test the pH of the water inside the actual growing medium by using a syringe or the likes, as sometimes the level of the pH in the growing area of the roots can be different

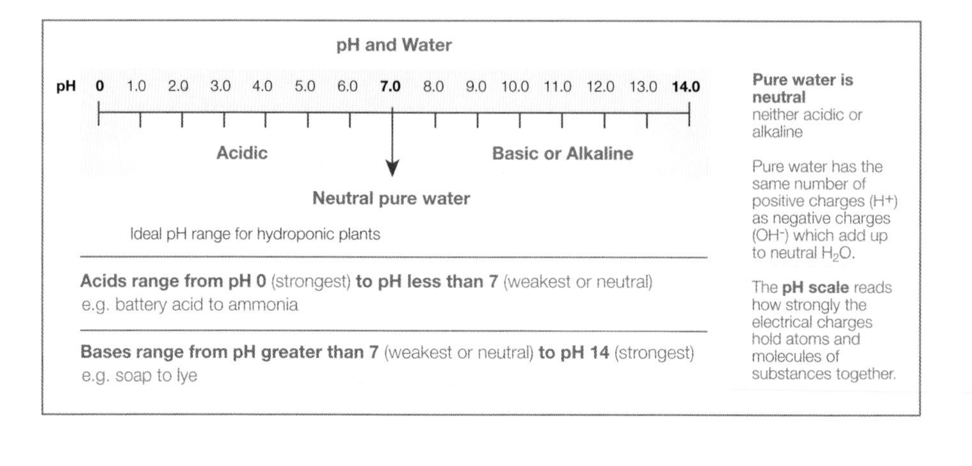

to that in the reservoir. If you find that it is different, then you should counterbalance this difference by adjusting the reservoir accordingly. That is to say, if you find that the solution inside the rockwool for example has a pH of 6.5 however, your reservoir's pH is reading 6, and you actually want the water inside the rockwool to read 6, then you should adjust the reservoir down to 5.5 in order to achieve this.

In most hydroponics systems and during most phases of growth, pH typically rises in your reservoir over the course of 2-7 days. So, you always need to readjust the levels of your reservoir. With this in mind, it is also advisable to allow a good pH swing in your reservoir, for example, if you want to maintain the reservoir's pH level at 6 to 6.5 then you can easily afford to lower the pH to 5.8 and allow it to swing all the way to 6.7 before knocking it back to 6. This swing encourages the plant to utilise different nutrients and actually strengthens the plant's immunity to stress and diseases. Therefore, it is not recommended to change the pH unless it actually needs it. The moral of this story, once again, is don't be anal – your plants will respect you more for it.

pH and Temperature

One word of warning: the temperature of your water affects the reading of your digital pH meter. A story springs to mind where a grower, for the life of him, was not getting the performance he should have been, as the plants were in a constant state of ill health, pale in colour, no vigour and very slow in growing. After hours of cross-questioning him

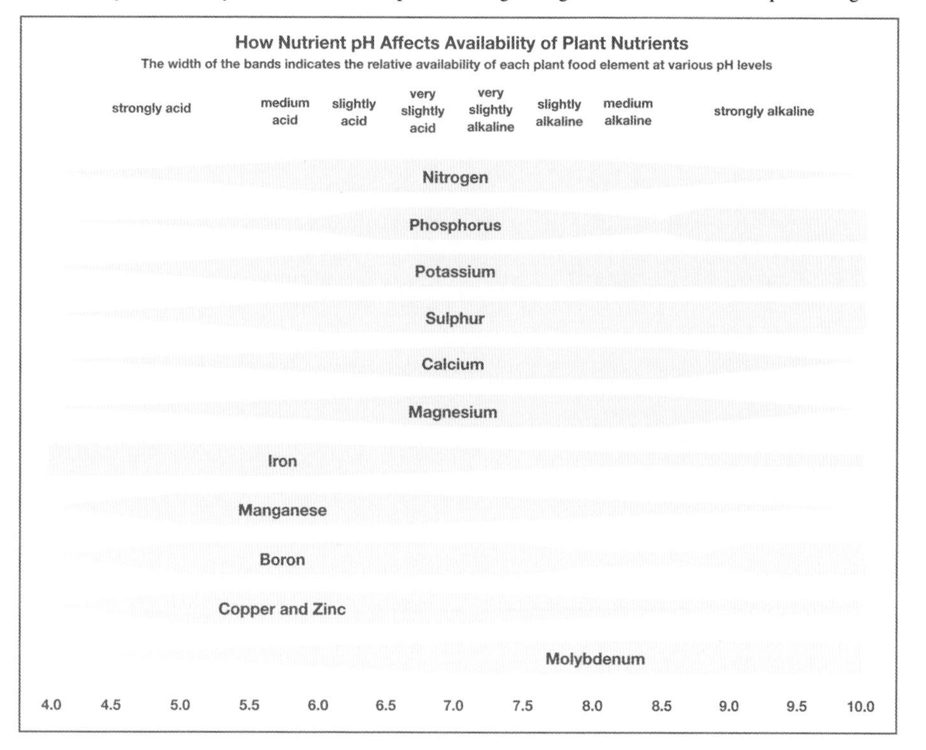

on all levels of running his grow room and feeding his plants and so on and so forth, at the end of the conversation as a passing note when he was leaving, he made comment that "oh, by the way, I always add a kettle of boiling water into the reservoir after filling it up." He delivered this statement as if every grower in the world followed this practice. So I then asked him " I take it you then balance your CF and pH after this?" Here was his blinding mistake. Directly after administrating boiling hot water to the reservoir he would adjust his CF and pH and in so doing, the reservoir having had a volume of very hot water inside it, therefore gave CF and the pH readings that were completely wrong. He would have been better off guessing rather than measuring it with his meters, as the volume of hot water changes the pH reading on your digital meters dramatically and also, but not to the same level, it affects the CF readings too. After showing him the error of his ways, the plants soon picked up and the grow was a good one.

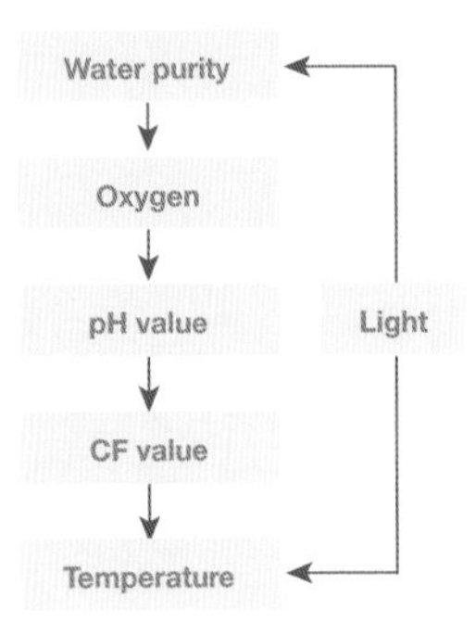

Basic Requirements of Plants

Hard or Soft Water

To know what type of water you are using when using a hydroponics technique is fundamental to the upkeep of your system.

The majority of nutrient manufacturers make hard water formulations of their products as well as universal and soft water formulations. The differing formulations will have differing buffering levels of pH stability; i.e. the hard water variations will reduce the pH in the water after dilution where the soft water variants will not affect the pH at all. This is beneficial to the grower because if you are in a hard water area when using hard water nutrients, it will lower the pH of your stock solution without you having to employ much pH down.

So the type of water you have is an important factor when considering what nutrient solution to purchase. Water quality does also play a major role in the quality of your end product. The worse your water quality is, then the harder the time the plants will have using it. In some cases of extremely hard water areas, it is highly advisable to employ a reverse osmosis machine to clean up the water before filling your reservoirs. These machines, by methods of complex filtration and run off, strip the water bare and in effect you get almost distilled water by removing all impurities from your water supply. If you are going to employ a reverse osmosis machine, it is still advisable to fill the reservoir with two thirds to three quarters using reverse osmosis (RO) water and one third to one quarter normal tap water. The reasons for doing this is that although very hard tap water has a high CF level and the majority of these dead salts are not beneficial to the plants, there are however, many trace elements and antibacterial agents that are valuable in many ways to the plants, so to add some hard water to the reservoir makes for good practice for keeping your plants healthy and strong.

As you can see, the type of water you use is an important bit of knowledge to have. Below is a breakdown of the pH and CF of hard and soft water.

Very soft water has a pH range of 5.5-6.5 and a

CF range of 1

Soft water has a pH range of 6.5-7.0 and a CF range of 2-4

Average water has a pH range of 7.0-7.5 and a CF range of 5

Hard water has a pH range of 7.5-8.0 and a CF range of 6-8

Very hard water has a pH range of 8.0+ and a CF range of 9+

Please note that these figures are approximations based on the UK water supply. Also, please be aware that water taken from your tap will vary on a week to week or month to month basis. This is due to the water companies using differing elements during different stages throughout the year to maintain the freshness and quality of water you expect. So, test your water from your supply regularly as the pH and CF do fluctuate depending on who supplies it, and when your water is supplied to you.

Tap water is the most practicable and inexpensive form of water supply to use. Rainwater is possibly cheaper however, if left for long periods of time can very easily stagnate and cause bacterial problems. Spring or mineral water is fine, however, very expensive and impractical unless your water supply is fed via a natural spring.

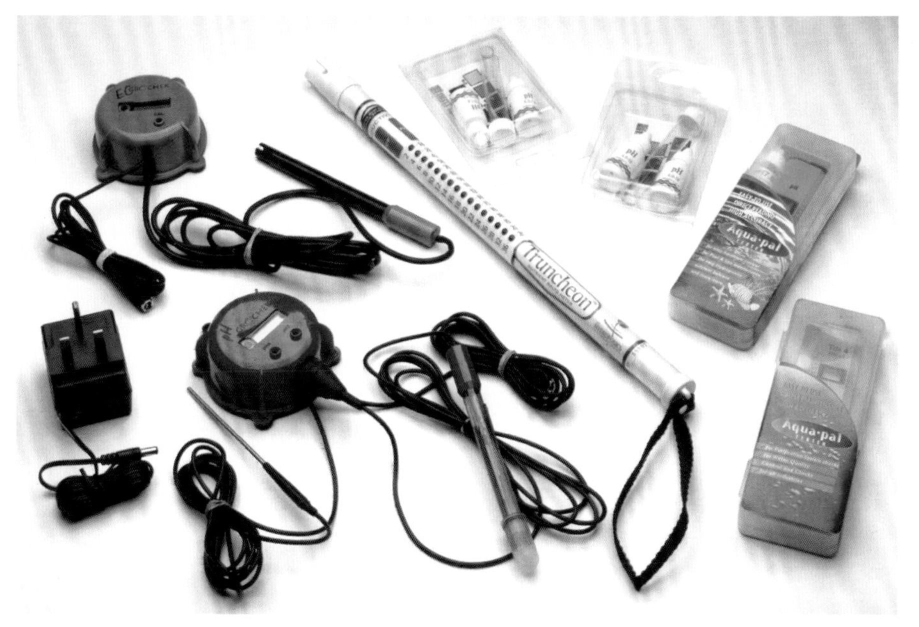

Examples of CF and pH Meters

Hydroponics
Indoor Horticulture

Chapter 9

Pests and Pest Control

Biological Pest Control

Recently, the indoor gardener has shied away from using chemical methods of pest control. There has been a huge range of biological control agents available to the commercial grower for many years and these are now becoming available to the indoor market in a readily usable form.

Biological control, otherwise known as predator bugs, are simply insects that feed on the insects that are feeding on your plants. These little bugs are not interested in your plants, they're only interested in the pests they can feed on.

In order to understand how biological pest control works, a thorough understanding of how the pests operate and function is required.

The life cycle of an insect from the time it emerges from an egg, to the time it lays eggs, may go through a number of definitive phases of body shape and behaviour.

Usually, the hatchling closely resembles its parent, although it may not have wings or may be a different colour. As the young insect develops, its hard outer covering will be shed several times and new characteristics will develop until it reaches adulthood and is ready to mate or lay eggs. Young insects are usually called nymphs, and an example of this type of insect is the thrip. Mites on the other hand are not insects but related to spiders and tend to follow their growth pattern.

Of course, there is always diversification. Some insects carry eggs inside until they are ready to hatch and so give birth to live young, for example, aphids. Others are very active during the first stage and are called crawlers. These scurry all over the place trying to locate appropriate places to feed prior to establishing themselves and becoming immobile. At this point, they become a miniature version of their somewhat immobile parent. This is the common scenario for mealybugs and scale insects.

Insects that have distinct young phases are a more complex situation. These insects go into adulthood via a transitional stage called metamorphosis (when their shape changes). The young are called larva (plural larvae) but during this intermediate stage they are called pupa (plural pupae) and in adulthood are sometimes called an imago. Good examples of these types of insects are butterflies and moths, wasps and bees, as well as flies and beetles. Common names are given to these stages which are different for each group of insects, namely, the larvae of moths and butterflies are called caterpillars, young flies are called maggots, and young beetles are called grubs. The intermediate stage of butterflies is often called a chrysalis and some insects will protect this perilous stage by creating a cocoon made of silk, while others might tunnel into the soil. Pupae may adhere themselves to a firm structure such as a wall, fence or plant.

Behaviour and appearance differs between these phases. The young are designed to be proficient feeders whilst the adults are most efficient at reproduction. So, on the whole, it is usually the one stage that causes all the damage to crops, for example caterpillars. The adult butterfly or moth does not do any damage while it feeds on nectar.

Two categories of insects are predominantly used as biological control – parasites and predators. Usually, a parasite will only destroy one host insect during its life cycle (egg to adult). Most often, they will feed inside the host while it is alive, and will only kill it when it is fully matured and ready to emerge. Predators, however, will kill scores of insects during their life cycle. On the whole, it is the adult that is the predator, although in some instances, only the larva is predatory.

Biological control is not a recent innovation, nature has obviously employed this technique since time began. In reality, it is a commonplace occurrence that happens everywhere in nature – in our gardens, hedgerows, woods and even in our homes. It is the natural cycle of predators eating prey. However, we now grow thousands of plant species from all over the world in our gardens, glasshouses and grow rooms, and as well as importing new plants, we have also imported many of the insects which would have lived on these plants in their native habitat. These insects are normally kept under control by predators, parasites and disease, creating a balance, so when these plants are cultivated, sometimes at great distances from where they originate, a complex and integrated system is left behind. These insects are no longer kept under control and are therefore able to multiply at a huge rate, damaging the plants, and so becoming pests.

Native species have also suffered by the indiscriminate use of chemical biological control. These pesticides are not particular as to which insects they destroy, so both pests and beneficial insects are wiped out. Initially, the insecticide seems to work, however, pest species usually recover rapidly compared to predators and parasites, so numbers are quickly re-established and before long are at unmanageable quantities. Once again, the chemical biological control is employed, but this time the pests are more resistant to the pesticide and may recover even more quickly. A few more cycles and the pest is almost immune to the chemicals and even worse, its natural predators have all been eradicated giving the pests free rein to do their worst.

In both these scenarios, biological control can be of use by attempting to restore the inherent enemies of the pest, so that it can be brought under control again. Many insects that have become pests are not indigenous to this part of the world, so insects have been imported to control them. This can be quite complicated as foreign insects cannot just be brought in without knowing what the impact might be to our own indigenous species. Biological control can only work provided the beneficial insects don't become pests themselves and prey on harmless insects or plants. A predator with an extensive range of prey may become problematic if it starts eliminating harmless native species as opposed to the pest species it was introduced to deal with. Therefore, prospective biological control agents undergo strict quarantine procedures and tests before being introduced to this country.

The majority of biological control is specific to the pest they deal with. However, if you decide to embark on biological control to control pests, insecticides cannot be used to control other pests, so a number of beneficial insects may need to be introduced over time.

As a method, biological control is a direct contrast to pesticides as results are not instantaneous. Beneficial insects need to establish themselves and increase their numbers before they can cutback the pest population. Biological control is hardly ever an immediate solution as it can take weeks, or even months, to manage the pest population

Environmental conditions are paramount when using biological control because of the use of live insects. Temperature, light and humidity will affect the outcome and at times, biological control will be unable to establish itself at all, so the grower needs to observe the situation very carefully. Biological control is not always a 'one-off' solution. Once the insects are established, they can keep control of the pests for a whole season, keeping pest numbers under control by augmenting their own population. Most times, there is a harmonious balance by keeping a small pest population that is adequate to maintain their food supply. However, in smaller areas such as grow rooms, it may be that the beneficial insects are too successful and may well completely eliminate the pest. This might sound great at first, but it also means that the beneficial insects will die because their food source has disappeared. If another outbreak should manifest itself later on, biological control will have to be released into the grow room again.

Most biological control is only appropriate in glasshouses, conservatories or grow rooms because many insects are from more temperate climates than ours, and can only become established in the warm temperatures found therein. Keeping them in these environments also solves another problem which is dispersal. If flying insects are introduced into the garden, they would simply fly away and the benefits will be achieved elsewhere. In a grow room, they are confined and therefore their effect is concentrated. As always, there are exceptions to the rule as in the case of biological control of caterpillars which can be used outdoors, as can the nematode treatment for vine weevils, sciarids and latterly, slugs. The red spider mite predator, *phytoseiulus*, can sometimes also be used outdoors. Some controls, such as *encarsia* (for whitefly), establish more quickly if small releases are evenly distributed over a number of weeks rather than in one huge dose. Something to note is that dealing with live products can be unpredictable, so with less common insects or those that are difficult to rear, it might not always be possible to source the products when required.

If identification is your thing, a good quality x10 hand lens will enable you to spot characteristic marks on insects as small as spider mites.

This next section consists of different descriptions of the major pests that can be dealt with using biological controls that are currently available, life cycles of the pests, symptoms of the types of plants that are being attacked, and the applicable biological control plus advice on how to use them to their best advantage. This is by no means a complete guide to growing plants without the use of pesticides. It should also be noted that some of the less common biological controls that are not yet available to the amateur grower, have not been included. Also, in the following sections, the term "grow room" includes any indoor growing environment, be it glasshouse or greenhouse, conservatory, polytunnel, loft, basement, spare room, shed or even your closet.

Aphids (Greenfly and Blackfly)

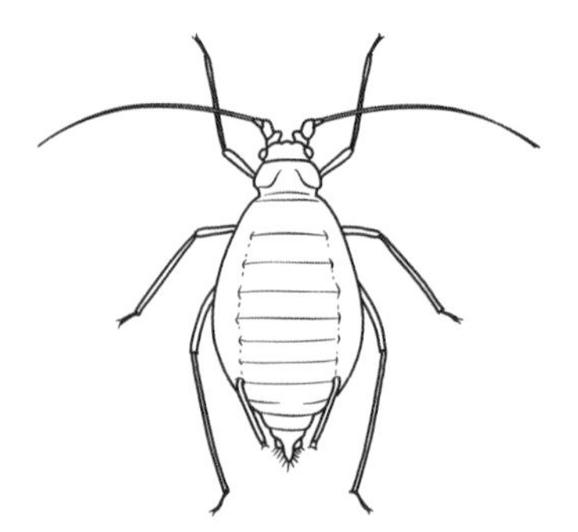

Example of an Adult Aphid

These small insects are soft bodied and pear-shaped and live in tightly packed colonies on plants. They come in a variety of colours and depending on the species and time of year, can be pale green, yellow-green, dark green, purple, brown or black. They have slender antennae and have two tubular projections at the rear of the body which are called cornicles. They can be either winged or wingless species. Sometimes, the cast-off skins of the aphids, which are white or translucent, adhere to the plants in and around the colony and these can sometimes be confused with whitefly.

Preferred Host Plants

Some types of aphids are very specific to the vegetation they consume, and some feed off a wide variety of plants. Two species of note are *myzus persicae* which is light green in colour and *aphis gossypii* – dark green in colour, attack a large variety of plants which can include tomatoes, cucumbers and chrysanthemums, C3 plants, as well as ornamentals.

Diagnosis

Aphids suck the sugary sap from plants by pushing stylets into the plant tissue. They excrete most of the sugar, which is called honeydew, and this covers the plants and adjoining areas producing a syrupy layer on which black sooty moulds grow. They favour feeding at the shoot tips so plant growth can be hindered and become misshapen. If the infestation is severe, the leaves yellow and the plants lose their leaves.

Life Cycle

Once aphids have overrun a new habitat, they can quickly increase their population and it has been estimated that a single aphid, provided her young survive, can produce millions of new aphids in one season. In indoor growing environments, all aphids are female, and produce live young which start feeding immediately. Newly hatched aphids already have young developing inside her, and can give birth within a week of being born. Young aphids are born at a rate of 3-6 a day over a number of weeks. When it gets a bit crowded, some aphids wander off to find less populated plants. Winged aphids are also hatched in over populated colonies, and they fly off and begin new infestations elsewhere. Aphids can overwinter on crops.

Biological Pest Control

Aphidius, Aphidoletes

Aphidius is a small, black parasite with slender wings measuring about 2 mm in length. It is only probable that the adult insect will be seen or the consequence of it parasitising an aphid. The parasitised aphid seems larger than normal and has a 'mummified' appearance. It comes in several different colours, contingent on the type of *aphidius,* which varies from a metallic gold or bronze, to brown or black.

Aphidoletes is a small and slender midge measuring 2 mm in length and has long legs. The males possess long grey antennae whereas those of the female are shorter and darker in colour. Usually visible are the larvae which are orange or orange red grubs that resemble maggots. When fully grown, these are 3 mm in length and the body narrows at both ends. They can be found in aphid colonies.

Life Cycle

Aphidius is a parasite and lays an egg into an undeveloped aphid. This happens very quickly and takes just a few seconds. The egg hatches after enlarging quite significantly as the larva matures inside the aphid. The aphid is destroyed when the larva reaches adulthood. When the larva is fully grown, which is very obvious, as the aphid becomes bloated and typically turns a metallic brown colour. The dead aphid is anchored to the leaf with a silk thread by the parasite and the larva spins a cocoon inside the aphid. It eventually emerges from the cocoon as an adult wasp via a circular hole which is visible in the back of the aphid. The entire process takes approximately 3 weeks at 20°C and a single female parasite can produce up to 100 eggs in her lifetime.

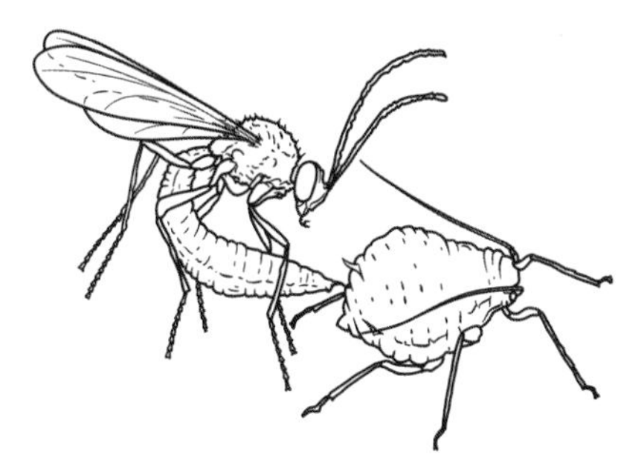

Example of *Aphidius* Parasite

Aphidoletes is a midge which produces a predatory larva. These feed on aphids in order to develop and reach adulthood. The adult midges subsist on honeydew from plants that are infested by aphids, and are active in the early evenings and throughout the night. During daylight hours, they are immobile and stay in the shade. Eggs are laid on the underside of leaves, but only on aphid infested plants, and they can lay about 100 eggs during their relatively short lifespan which lasts approximately a week. The eggs measure between 0.3 mm by 0.1 mm and are an orange colour and after a couple of days, small orange coloured larvae are hatched.

The larvae are voracious and must start eating aphids within a few hours of hatching. They can detect aphids over fairly large distances. *Aphidoletes* larvae poison aphids, which paralyses them before they die. This poison liquifies the body contents of the aphid after death and the larvae can then suck the body fluids out of the aphid. More aphids are destroyed than actually consumed, particularly if the aphids are big. The fully matured larva then makes its way down the plant or falls to the ground and then tunnels 2-3 mm into the soil, where it spins a cocoon of sticky threads coated in soil debris. The larva starts to pupate after a few days. If the day is less than 15 hours long, the larvae do not pupate straightaway, but remain in a state of suspended animation. In nature, this would occur in September, and the larva would then complete their development in the spring.

Procedure

Aphidius is usually delivered either as an adult wasp or parasitised (mummified) aphid, or possibly both. The tube is left open at the infested location, out of direct light and away from dripping water, so that the hatching parasites find their way out. Being a parasite they work better on small, isolated patches of aphids rather than huge colonies, so as soon as aphids are spotted, *aphidius* should be introduced. If aphids are prolific, numbers need to be cutdown with a soft soap spray, a suitable insecticide, or even plain water. *Aphidius* seem more productive on certain plants and seem to prefer smooth leaves. It is quite an able flyer, so any vents or doors that are left open should be netted over. *Aphidius* is productive under cloches if these are escape proof.

Aphidoletes is usually sent out in pupal form and mixed with vermiculite. The contents of the tube are scattered at the foot of infested plants, and caution should be exercised when watering so as not to wash the vermiculite away. The larvae are predatory, so handle large infestations of aphids quite well. However, some facets of its life cycle can make it awkward to use in particular situations. Firstly, it must pupate in the soil below the plant where it existed as a larva. If plants are grown in pots or grow bags, and the floor beneath is concrete or solid, the larva is unlikely to be able to find appropriate places to burrow and therefore, won't pupate and the next generation will be substantially diminished. Secondly, the larva tend to go into a state of suspended animation or diapause if they are released prematurely and will wait for several months before emerging as adults, by which time the damage has already been done by the aphids.

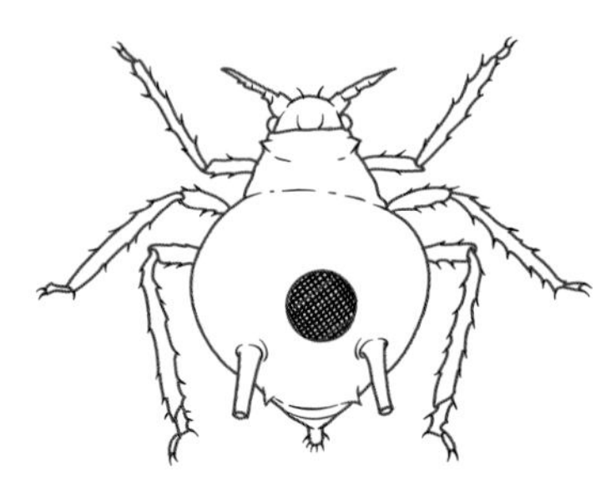

Example of a Mummified Aphid with Emergence Hole

NB – Biological control of aphids in the grow room can also be boosted by introducing native predators and parasites inside. The most familiar aphid predator is probably the ladybird and there are a number of indigenous species of this beetle, and most will consume aphids. Their larvae are avid predators. Ladybirds can be placed into the grow room whenever they are found and placed near to the aphids. They probably won't hang about for too long, but hopefully long enough to lay some eggs so that their young can feast on the aphids.

Another valuable predator is the hoverfly. The adult flies are usually enticed into the grow room by flowers, crops or ornamentals. The flies are actually nectar feeders but their young are aphid eaters. They look like short, oval-shaped caterpillars.

Another native is the lacewing. The adults are a bright green colour and have quite large gossamer wings. Large populations can be found in sheds and garages in the spring after overwintering in the garden. The larvae are often found around aphid colonies. They are brown with long, but paltry hairs on their bodies.

Aphidius itself is indigenous, and can often be found in the garden as parasitised aphids. If the mummies are undamaged and no exit hole is visible where the parasite has emerged from, they are worth collecting and placing in the grow room.

Fungus Flies (Sciarids)

Adult sciarids are 3-4 mm in length and resemble black midge-like flies. They can be seen hopping or hovering over the soil surface. Young sciarids are small white maggots with black heads, are 4-6 mm in length and are visible in the first few millimetres of soil.

Example of an Adult Sciarid Fly
(aka Fungus Gnat or Fly)

Preferred Host Plants

The flies prefer rotting vegetable matter including peat compost, especially if it's damp. When the population has increased sufficiently, they will attack living plants, especially seedlings, and all kinds of rooted cuttings and cacti.

Diagnosis

Seedlings are eaten at the base of the stem, at or just below soil level. Roots of cuttings are consumed, and damaged plants may rot.

Life Cycle

Females are attracted to the stench of rotting vegetable matter, including damp, warm, peat compost, particularly if algae is growing on it. The eggs are laid on the upper layer of the soil, and despite adults having a short lifespan, one female can produce between 100-300 eggs a week. The maggots hatch and eat the rotting vegetable matter and sometimes seedlings and cuttings. The entire cycle lasts about 4 weeks at 20°C and sciarids can infest grow rooms at any time of the year. They can also be present outdoors, so reinfestation of clean grow rooms can occur at any time.

Biological Pest Control

Nematodes

These tiny eelworms are supplied in a carrier medium mixed with water to form a suspension. They are only visible by magnification.

Life Cycle

The nematodes swim in the moisture that envelops the growing medium. They search for the sciarid larvae and gain access into them via body apertures. The intestines of the nematode carries bacteria and when these are unleashed inside the maggot, death occurs within 48 hours. The nematodes then breed inside the dead maggot, thereby releasing more nematodes into the growing medium.

Procedure

The nematodes and carrier material suspension are diluted and watered onto the surfaces of the affected areas. The growing medium must be kept damp in order for the nematodes to move around in the soil. Treatments to growing media may need to be administered frequently, that is to say, every 6-8 weeks to ensure the population does not increase to damaging proportions again. Nematodes can be stored for short periods of up to 4 weeks in a refrigerator, but will perish if frozen or dried out. This is a harmless biological control to other beneficial insects, to plants, humans and other animals. To ensure that infestation by sciarids is kept to a minimum, compost should be kept slightly drier and any dead plant matter should be disposed of on a regular basis.

NB – Nematodes as a biological control was

originally developed at the Glasshouse Crops Research Institute now called HRI Littlehampton, the very same place where Nutrient Film Technique was created.

Sciarids can be a problem to amateur growers but their economic influence as a pest was first felt in the mushroom industry where the sciarid fly flourishes in the environment where mushrooms grow. Larvae gnaw on the growing mushrooms and disfigure them, which makes the mushrooms unsuitable for sale. Until the nematode treatment was developed, pesticides were the order of the day. It was at this time that researchers discovered that the nematode would also attack soil dwelling larvae of beetles e.g. the vine weevil, and so nematodes were also aimed at this market too. Subsequently, *heterorhabditis* was found to be even more efficient against weevil larvae.

Another form of control on sciarids are tiny mites called *hypoaspis miles*. These creatures are soil and grow medium dwelling and grow to about 1 mm long. They will consume many small soil and grow medium living bugs, including sciarid larvae. *Hypoaspis* comes in a mixture of peat and vermiculite which is then administered to the grow medium surface.

Example of Mite *Hypoaspis*

Tomato and Chrysanthemum Leaf Miners

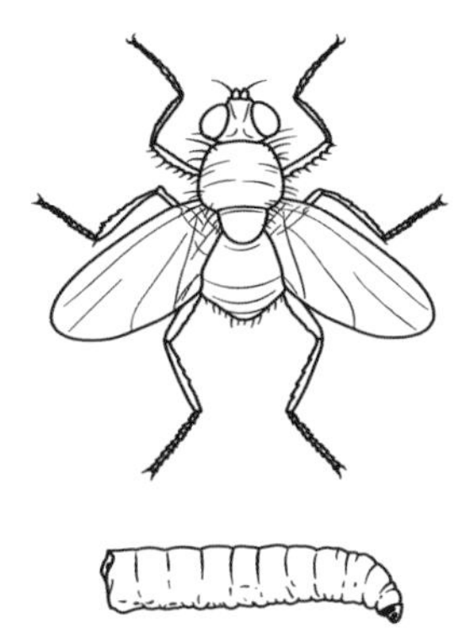

Example of a Leaf Miner Fly

The tomato leaf miner is a small black fly, approximately 2 mm in length with yellow spots on the body located between the wings. The chrysanthemum leaf miner is a grey coloured fly, approximately 3 mm in length.

Preferred Host Plants

The tomato leaf miner mainly attacks C3 plants, glasshouse vegetables such as tomato and cucumber, and some ornamentals. They can overwinter outdoors on sow thistle.

Diagnosis

Small white circles appear in the leaf tissue which are caused by ruptures made by the feeding adult female. White tunnels in the leaf tissue indicate where the larvae have burrowed. In acute cases, the leaves are so mutilated that the crops suffer, and in the case of display flowers, leaf mines are unattractive.

Life Cycle

Adult females choose suitable sites to lay their eggs while feeding. When the adult female makes a puncture, the egg is laid inside one of the small holes within the leaf tissue. When the larva hatches, it eats the internal tissue of the leaf between the upper and lower epidermis, thereby producing the familiar white, coiled tunnels on the leaf. The fully developed tomato leaf miner then drops to the ground where it pupates. The chrysanthemum leaf miner will stay in the leaf and pupate there, and can be seen at the end of its tunnel. The life cycle takes approximately 25 days at an average temperature of 20°C. Each female can lay about 60 eggs in 2 or 3 weeks.

Biological Pest Control

Dacnus Sibirica and *Diglyphus Isaea*

Dacnusa is a small black wasp approximately 3 mm long. *Diglyphus* is a tiny metallic green insect approximately 1-2 mm in length and has short antennae.

Life Cycle

Dacnusa is a parasite and lays its egg via the leaf tissue and straight into the miner larva in its tunnel. The *dacnusa* larva will develop within the live leaf miner larva only destroying it when the leaf miner pupates. Each parasite can strike about 100 leaf miners in approximately 2 weeks and the lifespan is about 3 weeks at 20°C.

Diglyphus females will attack any stage of the leaf miner larvae; she kills them by stinging them using her ovipositor, or egg laying tube. The egg is then laid alongside the body rather than in it, so that when the larva hatches out, it feeds on the dead leaf miner. As well as killing leaf miner larvae for their eggs, *diglyphus* will also kill other leaf miners in order to feed themselves. Approximately 60 eggs can be laid in 2 weeks and the lifespan is about 17 days at 20°C.

Procedure

Generally, a mix of *dacnusa* and *diglyphus* are distributed as live adults, as the effect of the two types of predator work in harmony. *Dacnusa* are good at finding small numbers of leaf miner, whereas *diglyphus* are more useful in heavily infested areas, killing the leaf miner straightaway thereby decreasing the number of tunnels made by the pests. *Diglyphus* prefer hot environments, and they will prey on leaf miner larvae that have already been parasitised by *dacnusa*, so it can become dominant in these environments. Parasites are introduced by placing the tube in the infested area and removing the lid. The tube needs to be placed out of direct light and away from dripping water. The insects find their way out and onto the plants.

NB – The two leaf miners mentioned here are flies and are related to the common housefly. However, many other types of insect exist that generate leaf mines. Many tiny moths (the *microlepidoptera* group) have caterpillars that develop inside leaves and produce mines, and there are also beetles (*coleoptera*) that produce leaf mines, and wasps (*hymenoptera*). Unfortunately, *dacnusa* and *diglyphus*, will only control the fly leaf miners.

Mealybug

Mealybugs are small, are soft bodied and have sucking mouthparts. They are from the same family as aphids, whitefly and scale insects. Usually, they are covered with a white waxy substance. Three species of mealybug can be found in indoor environments, although 15 species

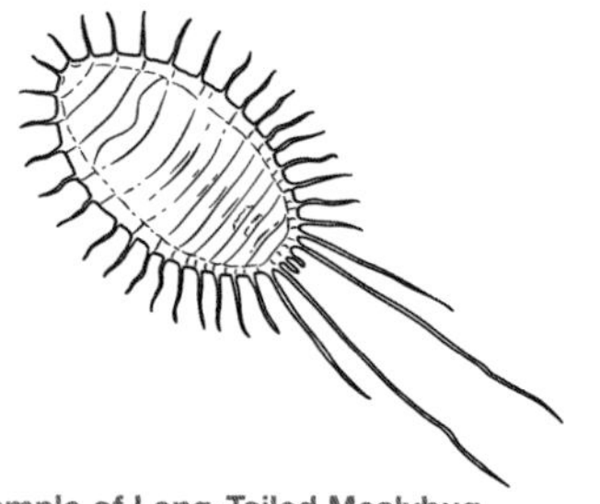

Example of Long-Tailed Mealybug
(*Pseudococcus Longispinus*)

have been documented. The citrus mealybug (*planococcus citri*), the vine mealybug (*pseudococcus affinis*) and the long-tailed mealybug (*psuedococcus longispinus*). They are recognisable by size, shape, colour and number of filaments on their bodies.

Females are oval in shape and up to 5 mm long. They are white or whitish-pink and filaments are visible around the extremities of the body and tail. If the species does produce males, they have a short lifespan and are distinguished by their delicate wings. The tail filaments of *psuedococcus longispinus* are as long as its body, and *p.affinis* possess a pair of short tail filaments.

Some species generate waxy wool in which to lay eggs. Most species will feed on the top part of the plant, but root feeders can exist and certain species cause abnormal growth in plants.

Preferred Host Plants

Planococcus citri has been recorded on 25 different plant groups and other species can be found on most ornamentals and C3 plants. *P. affinis* is the most commonly recorded species on cacti, passiflora and tomatoes.

Diagnosis

Female mealybugs usually congregate, often in leaf whorls or along the stems. All species produce honeydew which covers the plant and neighbouring areas and leads to black, powdery mould growth which is unsightly and in more extreme cases can deprive the leaves of light. Large populations of mealybugs can weaken the plant because so much sap is being taken, which might cause leaves to yellow and even loss of leaves. Root feeders appear as waxy, white patches among the roots and are usually seen when repotting plants, especially between the roots and pot sides.

Life Cycle

Most mealybugs lay eggs, although *p. longispinus* produce live young. *P. citri* and *p. affinis* produce woolly white masses of wax filaments in which they can lay up to 500 eggs. Egg laying can take up to 10 days, and the female mealybug significantly decreases in size. When the egg laying has finished, the female dies. Hatchlings are very mobile and are called crawlers. They scatter quickly to find suitable sites where they can settle and start feeding. Those species that produce males, pupate and delicate winged insects that look like whitefly come out. Females carry on feeding until they are able to lay eggs. The whole cycle takes approximately 50 days at 20°C or 25 days at 30°C, although warmer than 30°C will hinder egg production.

Biological Pest Control

Cryptolaemus Montouzieri and Parasitic Wasps

Cryptolaemus is a ladybird originating in Australia. It is a black or dark brown in colour and has an orange head and tail. It is about 4 mm in length.

The larva is white and resembles a very large mealybug, up to 1 cm long, and is covered in a wax-like material. It is more active than the mealybug.

Parasites that are most readily available are *leptomastix dactylopii* which are used to control the citrus mealybug and *leptomastix epona* which are used to control the vine mealybug. They are tiny wasps, approximately 3 mm in length, and colours range from yellow to black. Parasites that attack other species of mealybug are sometimes available and currently research is being done on parasites of the long-tailed mealybug.

Life Cycle

Cryptolaemus females start laying eggs about 5 days after reaching adulthood. They lay single eggs into the mealybug wool egg masses and can lay about 10 eggs a day and up to a maximum of 500 eggs. In order to produce eggs, adults need plenty of food – starving adults will not produce eggs. When the larvae hatch, they are extremely voracious predators on mealybugs. Young larvae and adults favour the young mealybug, but the large larvae will eat any size of mealybug. *Cryptolaemus* aren't fussy, and if mealybugs are in short supply, they will eat other insects such as young scales. The larvae can turn cannibalistic and will eat each other. This is why adult beetles are confined in tubes when sent through the post. The life cycle takes about 25 days at 30°C or 72 days at 18°C.

Leptomastix females will lay their eggs in almost adult mealybugs. The parasitised mealybug then swells up and turns brown in colour, prior to becoming 'mummified'. The parasite adults emerge by making a small circular aperture in one end of the "mummy". The life cycle takes about 4 weeks at 20°C and 2 weeks at 30°C. Eggs that have been laid into other species are destroyed by the immune system.

Procedure

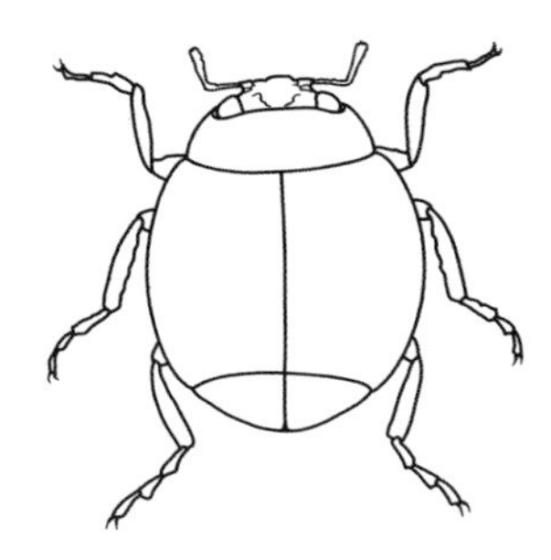

Example of Adult *Cryptolaemus* Beetle

Cryptolaemus usually come in the post as adult beetles. They should be released near to the infestation site and in the cooler part of the day, so that they're not grumpy and just fly off immediately. They are very able and strong fliers, so can escape through open vents and doors. This may necessitate the use of screens over these openings. *Cryptolaemus* prefer bushy or leafy plants rather than climbers and they are also averse to some plants with leaf hairs, especially irritants like those of the tomato. Because the woolly egg masses of the mealybug are preferable for laying eggs into, any species that don't produce these masses, may not be successfully eliminated. *Cryptolaemus* likes quite large populations of mealybug and so may never entirely eliminate the pest.

When it comes to cacti, due to the size of the beetle and its larva, this can prevent them from accessing the mealybugs that may be sheltering between ribs or under the tight-knit spines. Cacti are not favoured by *cryptolaemus* due to the physical layout, with each plant affording a minimal hunting area, and plants being arranged quite far apart. It may well be that *cryptolaemus* does not perceive cacti as plants anyway!

Cryptolaemus cannot get to root mealybugs and so are not able to control them and as a number of species can be both aerial and root feeders, they can be difficult to eradicate.

Leptomastix and *cryptolaemus* are usually applied in combination, but can be used independently. They are supplied as adult wasps and released near to the infested site. As referred to before, *leptomastix* only prey upon certain species of mealybug, so correct identification is required to ensure the right control is used. Temperatures above 25°C are needed for *leptomastix* to thrive. As *leptomastix* is a parasite, it is more proficient at searching out solitary patches of mealybug, so is effective when used with *cryptolaemus*. Employed in combination, they can eliminate infestations of citrus mealybug.

Cryptolaemus and *leptomastix* need sufficient levels of light to thrive and don't do particularly well in shady conditions, or in rooms whose only source of light is from a window. Because the parasitic wasps are specific to certain types of mealybug, correct identification is crucial prior to placing an order for the correct sort of biological control.

NB – Do *cryptolaemus* bite humans? They definitely bite mealybugs and like the native ladybird, have quite powerful mandibles. If they land on skin they have the ability to deliver quite a bite, but it is doubtful if this would perforate the skin. Their bite is not poisonous.

In the first part of this century, *cryptolaemus* was one of the first successful endeavours of pest control using predators. It was brought in from Australia to California and released among the citrus groves which had experienced significant destruction brought about by the citrus mealybug. Although it was very successful, and the beetle became known as the 'Mealybug Destroyer', it sometimes died in the Californian winter. Large scale production of *cryptolaemus* in insectaries was launched, so that large numbers could be released every year. And that is how the system of periodic colonisation was introduced to grow rooms.

Red Spider Mite (Two-Spotted Mite)

A small mite which is yellow or olive-coloured

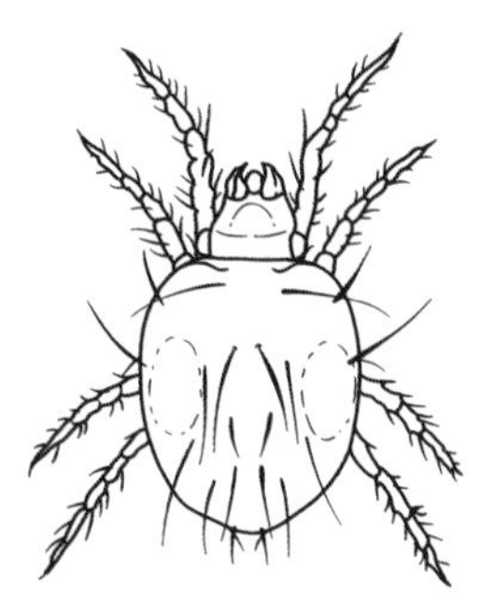

Example of Red Spider Mite

with dark patches on both sides of its body. They measure less than 1 mm in length and can appear in large numbers on the undersides of leaves. In due course, they create fine webs that cover the plants on which they walk. In autumn, the mites turn deep red in colour.

Preferred Host Plants

A wide range of C3 plants, glasshouse vegetables and ornamental plants.

Diagnosis

The mites extract the contents of the leaf cells which produces a mottling effect on the upper leaf surfaces. Heavy infestations produce the characteristic webs and if not controlled, the leaves go brown and drop off and the plant may die.

Life Cycle

The eggs are deposited on the underside of the leaf. There are five development stages and egg to adulthood takes about 14 days at 21°C, and less than a week at 30°C. Adult females can produce more than 100 eggs over 3 weeks. The reproduction rate from one generation to the next is approximately 31 times! In the autumn, when days are less than 13 hours long, the mites turn deep red and leave to hibernate in nooks and crannies in the glasshouse. They can overwinter without feeding, and emerge in the spring and summer to infest crops once again and with artificial lighting, they may only hibernate for very short periods of time (between crops), or they may not even hibernate at all. Humid environments that are higher than 60% RH can diminish the egg production of a mite.

Biological Pest Control

Phytoseiulus Persimilis

This is a predatory mite originally from Chile, and is somewhat larger than the two-spotted mite. It has a pear-shaped, shiny, red body, long legs and moves very quickly as it searches for its prey. The young are oval-shaped and a very pale pink colour. *Phytoseiulus* usually comes mixed with vermiculite carrier grains which are shaken over the infested crops.

Life Cycle

Each mite can lay between 50-60 eggs over a period of 3 weeks. Egg to adult takes about 12 days at 20°C and about a week at 30°C. Hundreds of spider mites can be eaten by a single predator during its life, and all stages of Red spider mite are acceptable to the predator. Once established, the predator can rapidly overwhelm the pest, which isn't difficult, as the mite population increases 44 times in one generation!! If all the pests are eliminated, the predator will scatter and die.

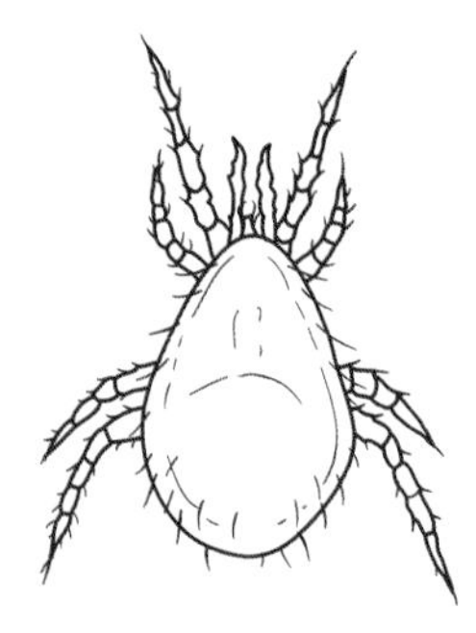

Example of *Phytoseiulus Persimilis*

Procedure

Phytoseiulus should be introduced as soon as mites are spotted. The air temperature must reach about 20°C for at least some of the day, and humidity needs to be kept quite high at above 60% RH by damping down or misting. Red spider mites aren't particularly fond of high humidity and misting in conjunction with the release of *phytoseiulus* will bring the mites under control more quickly. Sometimes, the predator is so successful and completely eradicates the spider mite, so it may need to be reintroduced if further infestations occur.

NB – There are several other mites that can be classed as pests in the grow room and even more that are harmless or even beneficial. A relative to the Red spider mite is the Carnation spider mite, *tetranychus cinnabarinus,* and as the name implies, often causes trouble with carnations. The Broad mite is tiny and almost colourless and can cause leaves on ornamentals to go brown. Its relative, the Cyclamen or Strawberry mite, causes leaves and buds to deform. *Phytoseiulus* is not particularly proficient at wiping out these mites. However, intensely red, fast moving mites that seem to have a pair of legs in front of them are predatory Trombid mites and are harmless.

Scale Insects (Soft Scale)

Scale insects that infest grow rooms can be divided into 2 main groups; the *lecanidae* (soft scales) and the *diaspdiae* (armoured scales). Here, we will only deal with the soft scales. *Coccus hesperidum* is oval-shaped, has a flattened appearance, and its colour ranges from green to brown. The adult is about 3-5 mm in length. Young scales are paler in colour and are yellow or greeny-yellow. They can often be seen along leaf veins on the undersides of leaves. *Saisettia coffeae* is larger and dome-shaped, and the adults are dark brown in colour.

Preferred Host Plants

There have been 37 or maybe more plant genus documented as hosts for soft scales and these include C3 plants, ferns, orchids and the majority of ornamental plants.

Diagnosis

Large numbers of scales may cause the plant to yellow and loss of leaves to the most affected branches. But more damaging is the huge amounts of honeydew that are secreted by the scales, especially *coccus hesperidum.* Plants and adjacent areas become sticky, and black powdery moulds soon grow on the sugar. This looks ugly and light reaching the leaves is severely restricted, thereby starving the plants.

Life Cycle

The life cycle of both these species is much alike, in that the adult females (there are no males) produce a horde of eggs under the scale cover over several days, and then die. *C. hesperidum* can lay up to 250 eggs, whereas *S. coffeae* can lay up to 2000 eggs. The eggs are less vulnerable to pesticides whilst beneath the protective scale cover. The hatchlings are called crawlers and are small legged creatures. They leave the scale cover and rapidly scatter all over the plants and nearby leaves. Their main priority is to find appropriate plants on which to install themselves and become immobile. Most crawlers die at this stage, but the survivors establish themselves and become the more familiar scales. Scales can usually be found along veins or stems where they can feed on the sugary sap which is carried in the phloem vessels of the plant. Scales have quite a long lifespan; the life cycle of *S. coffeae* takes approximately 95 days at 18°C. Soft scales flourish in temperatures of around 20°C but are unable to thrive in temperatures above 30°C.

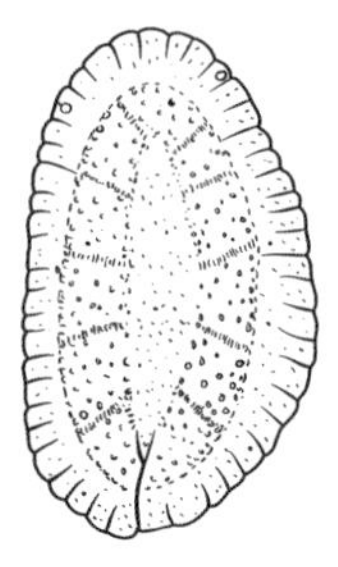

Example of Brown Scale
(Coccus Hesperidum)

Biological Pest Control

Metaphycus Helvolus

Metaphycus is a tiny parasitic wasp approximately 2 mm in length. Females are yellow in colour and are quite lively, moving rapidly on leaf surfaces and are capable of jumping.

Life Cycle

Adult females will attack young scales once they are established. A female can lay about 6 eggs a day into the scales, but kills maybe four times this number when testing with her ovipositor, or egg laying tube, and then feeding on injured scales. The larva grows inside the scale which is finally killed, and its appearance seems flatter and darker than normal. The *metaphycus* then pupates inside the scale and comes out fully matured by cutting a circular perforation at the top of the scale. This whole process takes about 30 days at 20°C and 11 days at 30°C. Adults have comparatively long lifespans, perhaps 8 weeks if adequate food supplies are available such as young scales or honeydew.

Procedure

Metaphycus are supplied as live adult wasps. They cannot survive without food or moisture, and must be released into the growing environment immediately. The tube is left open near the infested area (out of direct light) so the insects can escape. To be effective, there must be plenty of light in the grow room and the temperature must reach 22°C for at least a few hours in the day. As *metaphycus* prefers young scales, several releases of them might seem to be the correct route to take, but as the adult has a long lifespan and production time is short, a single release is all that is needed. In indoor growing environments, where the temperature remains constant during the winter and young scales are present, *metaphycus* can overwinter and re-colonise.

NB – Currently being researched is a beetle predator of the armoured scales (*diaspids*) and reports seem to be encouraging. It is called *chilocorus* and the necessary permission for its release in this country has been obtained. However, this will probably be of greater significance to indoor growers as temperatures of around 25°C are needed. There is currently more work on *diaspid* scale parasites in progress but they probably won't be available for some time.

Thrips (Thunderflies)

Thrips are small, slender insects which measure about 2-3 mm long when mature. On adults, two pairs of narrow wings fringed with long hairs are visible and when at rest, the wings are laid along the back. Their colour depends on the species, but these can range from a pale yellow through to a greyish yellow-brown to black. In order to establish characteristics which distinguish the different species, magnification is required. Recently, new species have been introduced from abroad and can be potentially damaging pests, such

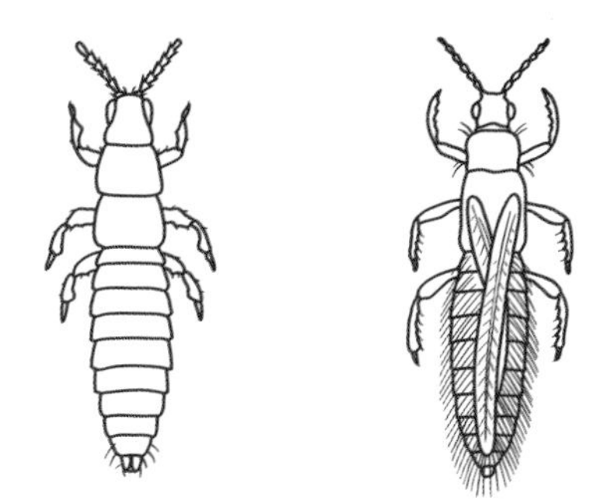

Examples of Thrips - Nymph (left) and Adult (right)

as Western Flower thrips, Glasshouse Banded thrips and Palm thrips. Thrips like company and usually appear in large numbers on leaves or flowers.

Preferred Host Plants

Many C3 plants, crops and ornamentals can succumb to infestation, most frequently on cucumbers and flowering plants.

Diagnosis

Thrips suck sap by puncturing leaves or flower buds with their mouthparts, and the affected parts become blotched or flecked and dry out. In very acute cases, the plants appear to have been burned. When leaves or buds swell, they can become contorted or torn, and form windows in the tissue. The thrips are visible, usually in large numbers, near the veins on the undersides of leaves.

Life Cycle

A female can produce approximately 60 eggs during the course of a season. Eggs are deposited on the undersides of leaves, protruding from a tiny cut in the plant tissues. The larval stage is about 10-14 days in the case of the common Tobacco thrips, whereupon the larva drops to the ground and tunnels into the growing medium to pupate there. Reaching the adult stage takes another 4-7 days. Thrips are quite vulnerable to insecticides, but can be quite awkward when other biological control agents have been used and this prevents chemical control. Fumigation between crops can be a useful remedy, but pesticide residues can remain for some considerable time and prohibit future use of biological control in the grow room.

Biological Pest Control

Amblyseius Mackenziei and *A. Cucumeris*

Both species of *amblyseius* are small predatory mites approximately 1 mm in length with slightly flattened bodies which are pear-shaped. They are a pale brown colour. They are very energetic and scatter over a wide area. Their white eggs can sometimes be seen attached to plant hairs.

Life Cycle

Amblyseius have quite a long lifespan, and can produce several eggs a day. Both adults and young mites will eat thrips. They can also eat bugs other than thrips, for example, young spider mites, so they can survive even if the thrips population has been diminished.

Procedure

Amblyseius are sent out as mobile mites that are mixed with bran. This is sprinkled over infested crops. Now available are small sachets that contain bran and mites which can be hung among the plants where the mites then scatter. These sachets also contain enough food for the *amblyseius* to survive should food be in short supply, or if they are introduced before the thrip population has multiplied enough. These sachets are sometimes called 'breeder' or 'grower' packs. *Amblyseius* are not particularly fond of low humidity, and lengthy, dry and hot spells can cause their numbers to reduce quite significantly. Damping down or misting during these hot periods can be beneficial. These sachets should be placed out of direct light and away from heat sources.

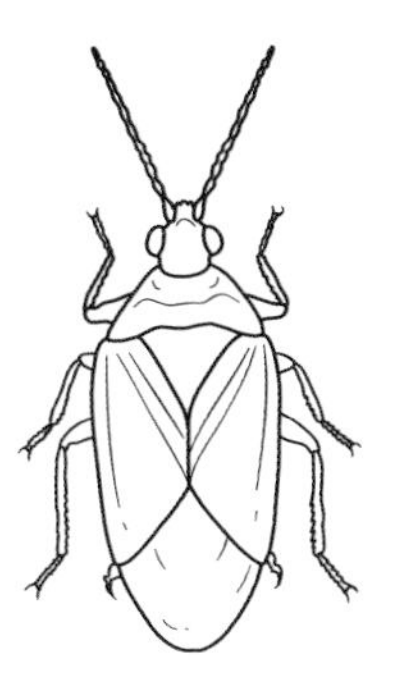

Example of Pirate Bug (*Orius Majusculus*)

NB – Recently, another predator called the *orius* or 'Pirate Bug', has been introduced to reduce thrips infestations. There are several species of this bug, of particular note is the *orius majusculus,* which is a voracious predator and will eat other bugs apart from thrips, including other *orius*! However, they are difficult to establish in the grow room and because of their cannibalistic nature, introducing them in any quantity is quite an obstacle. Another downside to the *orius* is that it may take a nip at the grower! Species of *orius* that are less malicious are being researched, but foreign species may not get the go ahead to be introduced to this country. *Orius* is usually available in commercial quantities, but there are reservations about its benefits in the amateur grow room. Research is currently being done on thrips parasites from overseas, but it is too soon to establish their viability as a biological control.

Whitefly

This is a small flying insect about 3 mm long, looks similar to a tiny moth and is white in colour. When not flying, they can be found on the undersides of leaves. Young stages look like small translucent scales on the undersides of leaves.

Preferred Host Plants

A wide range of crop, ornamental and C3 plants. Preferred plants may include cucumbers, fuchsias, pelargoniums, peppers and tomatoes, and many more.

Diagnosis

The young will suck sap from the plant and cover the leaves, pots and floor with sticky honeydew. Black powdery moulds grow on this sugary substance and envelop the plant in a black layer which limits light reaching the leaves. Extensive infestation may cause the plant to yellow, wither and die.

Life Cycle

Clean grow rooms, glasshouses and conservatories become infested when plants are brought into this environment. The eggs are deposited on the undersides of freshly grown leaves and the eggs hatch into small, round, scale-like young. These adhere themselves to the leaf where they stay immobile and feed on the sap. Scales will go through several phases before maturing into adults.

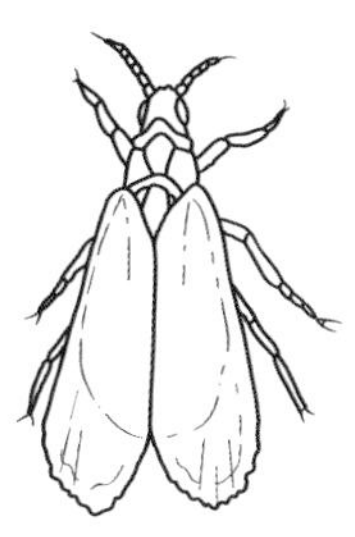

Example of Whitefly

Egg to adult takes approximately 30 days at an average temperature of 20°C. The lifespan of adults will depend on the plant that the young were reared on, some plants produce long-lived and fertile adults. Whitefly can survive winters on plants in the grow room, even on weeds, and eggs can withstand brief periods of frost. Many whitefly show significant immunity to chemical pesticides. It is a widely held belief that whitefly can migrate from infested allotments or vegetable gardens into indoor growing environments, but in truth the whitefly that infests outdoor vegetables is a very different species and does not infest grow rooms.

Biological Pest Control

Encarsia Formosa

Adults are tiny black and yellow wasps measuring 1 mm long. They usually come as parasitised whitefly scales and contain the *encarsia* in its pupal stage and resemble small black scales. Scales can be supplied in various formats, either loose in a plastic tube, stuck to sticky cards, or as parasitised scales on the host plant leaves (usually tobacco).

Life Cycle

Encarsia formosa attacks and kills the young immobile whitefly. Females are attracted to the scent of the honeydew and lay their eggs in the scales. During her short lifespan (10 -14 days) she can produce up to 60 eggs. Approximately 10 days after being parasitised, the scales turn black. Egg to adult takes about 28 days at an average temperature of 20°C, 14 days at 30°C. *Encarsia* activity depends very much on light and temperature and sunny days in March and subsequent months offer an ideal environment in the glasshouse, or all year round in an indoor grow room

Procedure

Encarsia should be introduced as soon as whitefly are spotted in your crops, and to be truly successful, should only be introduced if the whitefly population is not too large. This can be problematic in the earlier part of spring when temperatures and light are inadequate for *encarsia* to be successful, but the whitefly population that has successfully overwintered is already beginning to proliferate. Where the population of whitefly is already high, alternative methods need to be employed in order to reduce the whitefly population before the introduction of *encarsia*. One or all of the following may help to reduce the whitefly population –

- yellow sticky traps just above affected plants may be useful.
- spraying affected plants, especially new growth, repeatedly with water mixed with small amounts of soft soap or washing up liquid, including the undersides of the leaves.
- use a small vacuum cleaner as often as possible to remove adult whitefly around the plants.
- spray new and old growth with organic fungicidal soap solution, making sure the entire plant including the undersides of the leaves are sprayed.

If these methods are unsuccessful, using insecticides may be your last recourse; if you can find one that works on your whitefly infestation! Be aware that *encarsia* can be adversely affected by pesticide residues, so allow several weeks to elapse between spraying and introducing *encarsia*.

Research is currently underway on some foreign whitefly predators. These include a tiny black ladybird called *delphastus* and a small green predatory bug called *macrolophus*. These predators attack concentrated colonies of whitefly, but it may be a while before they are available on the market.

Organic Fungicidal Soaps

Brand new innovations in this area of pest control have recently propelled this method of control into the forefront of pest eradication. These types of organic pesticides are termed direct contact spray solutions. They are typically retailed as a concentrate, which is then in turn diluted down as per instructions on the bottle, then liberally applied by means of a hand held or pump action mister. Direct contact means that in order to kill the bugs you need to hit them with the solution; if the solution does not reach them then it will not kill them i.e. if the pest in question is sheltered under a leaf for example, then the pest will not be harmed at all and can reproduce at will. So, when using this type of product, it is very important to make sure that the entire plant, from top to bottom, is saturated with the spray. For that matter, some species of bugs jump when administrating the spray, so it is recommended to spray the entire grow room just to make sure that you have got them all!

This new type of product works by physically blocking the breathing holes of the insects and mites, quickly killing them. This product is naturally derived and indeed is also non-toxic. It does not leave behind any poisonous by-products or residues, unlike many commercially available pesticides which we strongly recommend not to use. The long term effects of industrial pesticides have yet to be charted or monitored, so if it is a poison, then keep clear from using it, as who knows what the long term side effects can be, both environmentally and domestically.

The new organic fungicidal soap solutions that are now available are safe to use on consumable end products, and as they are made from plant extracts themselves, they are also approved by organic gardeners.

These products have another important part to play in helping beneficial insects, predator insects and parasite insects, allowing these insects to continue their natural role in controlling pests. Most conventional pesticides leave behind poisonous residues, either on the plant or inside the plant, or in the local environment, and this aftermath can often harm beneficial insects for longer than the pest themselves, creating a large imbalance for nature to deal with. Typically, the pest over a period of time, comes back to infest with greater ferocity than before. However, the new fungicidal soaps will only kill insects that it comes into contact with before it dries. After this product is dry to the touch, it is no longer a threat to any living insect. Which means it makes an excellent head start on destroying pests before you unleash your predators to finish the job. As soon as it is dry, the predator bugs can be released. It is advisable to use the two in conjunction with each other as the fungicidal soaps do not harm the eggs of the pests. When using the two together, you basically wipe out the majority of the infestation with this product, then releasing the predators to deal with what is left, which dramatically increases the odds of the predators becoming dominant on release. The fact that these products are harmless when dry means that beneficial insects, whether introduced as biological control or naturally occurring, can carry on working almost without interruption.

These products in terms of safety for yourselves and your plants cannot be beaten. They have been tested on a large number of different plants with no problems. And with its reassuringly physical mode of action i.e. suffocation, it is highly unlikely that the pest will develop resistance to these products in the future.

These products are effective at controlling aphids (greenfly and blackfly), Red spider mite (two-spotted mite), whitefly, thrips (thunderflies), leafhopper, mealybug, and scale insects. They are highly effective, fast acting, easy to use, non-toxic, non-chemical, non-poisonous, safe to use with biological controls, tested on many plant species with no phytotoxicity problems, have no restrictions on how often they can be used, are cost effective and approved by organic growers. What more can we say, and if you are still using industrial pesticides, WHY?

Pests, if left unchecked, will simply multiply at an exponential rate. One can become some in a day, some can become many the next, and many can become uncontrollable in less than a week. If you do nothing, then most pests will decimate most crops in no time at all. If you get a pest problem, then it is highly advisable to literally wage war on your pests. Attack them, and the situation, from all angles. It is a very good idea to nuke the adult population repeatedly over 2-5 days with a good fungicidal soap like Eradicoat before sending the troops in to follow through. In so doing, this will give your predator bugs a good head start on the war ahead. With this in mind, this is an insect war that you have just instigated, so remember, if you have a large infestation of pests, then make sure that you send in some reinforcements to help finish the job.

Example of Eradicoat

Chapter 10
Fungi

To understand fungi and fungi problems associated with plants, you firstly have to understand what fungi are. Fungi are in fact very rudimentary forms of plants, however, they do not have to produce chlorophyll to survive. Fungi survive by reproducing through the spreading of tiny microscopic spores rather than seeds. Millions of these spores are typically present in the atmosphere at all times. In an affected grow room, this number is concentrated 100-1000 fold, creating a heavy soup of fungi spores awaiting to reproduce and procreate using your plants as their host. So when these microscopic airborne spores find suitable conditions, they simply settle on your plants, take hold and start to grow. Some fungi and as an example botrytis, are such prolific breeders that they can spread through an entire crop in a matter of days, so be warned, this stuff is treacherous.

Botrytis

Botrytis also otherwise known as bud rot or grey mould, is encouraged to grow in a grow room which is not properly ventilated. This means being properly ventilated at night as well as in the daytime. If your temperature and humidity are all pukka during the day cycle you should also make sure that the temperature and as importantly, the humidity during the night time is also controlled and ticker-dee-boo. One grower for the life of him could not work out why he was getting bud rot when during the day cycle the conditions were near on perfect. The night time temperature was also not bad however, he failed to recognise that as the fans were switched off at night, the temperature was keeping within the parameters, however, the humidity was not, so each night the humidity was rising to 98% and the walls were literally dripping with moisture. During the day this would even out and to all intents and purposes, looked good and well, but repeated heavy night time humidity soon took its toll on these plants. During the latter stages of flowering, botrytis took hold and then, and only then, realising his mistake, he cut out the infected flowers and increased extraction in the grow room during the night period. However, the damage was already done and this grower was forced to take down his crop one or two weeks prematurely resulting in a definitive lack of yield. Still, he did try and persevere, but with any fungi outbreak, you are fighting a losing battle and it becomes a matter of time versus damage, so you need to know when to draw the line to minimise casualties, but maximise yield.

Example of Botrytis

Prevention

Prevention before cure in the case of botrytis is key to controlling and eliminating the possibility of fungi outbreak. So common sense prevails when preparing a grow room. Remove all old, dusty and even new fabrics that could harbour fungi. If you cannot remove carpet for example, then cover it with a plastic membrane and this will protect your carpet from spills but also protect your grow room from possible fungi that may already be present inside the carpet. It is also advisable before installing your grow room to spray all and everything inside it with a diluted H_2O_2 solution to kill any possible spores that might exist in the air, on the walls or in some nook or cranny. After effectively nuking the grow room space, walls, floor and ceiling and general area, then mop up the excess moisture and allow the entire room to dry out.

After doing the above, cleanliness applied to you and your grow room are paramount. Followed with proper day and night time climate control, then very few clean and well ventilated grow rooms will have difficulties with fungi. On the flip side, most stale, dank, dirty and badly kept grow rooms develop fungal problems in one way or another, which in turn has a direct knock-on effect towards yield and productivity.

It is imperative to keep humidity down to around 50% day and night. This can be easily achieved by installing adequate extraction fans wired to humidistats and thermostats or by simply having your fan switching on every now and then during the night time cycle. If you are still faced with high humidity then very effective dehumidifiers are readily available from any general hardware store and are cheap to buy and run, effectively reducing excessive humidity inside your grow room. If temperature and humidity are a problem for your grow room and you have already employed extraction fans but are still having difficulties containing the heat and the humidity, then your next only feasible option is to use air conditioning. These machines will reduce the humidity and temperatures inside your grow rooms very efficiently, however, they are expensive to buy and run, but if all else fails they will do what they are meant to do. Experienced growers also have these machines available to fall back on if the situation dictates it.

So prevention is imperative and simply achieved by controlling all and any factors contributing to fungi growth or spread. If, however, prevention proves inadequate and fungi appear, then drastic measures are called for. Carefully remove and destroy dead leaves and infected fruits or flowers, wash your hands before and after handling the affected areas. If you can isolate affected plants then do so as soon as you can. Fungi, if left unchecked, and even in cases where you have addressed the problems behind the breakout, can still exponentially procreate and consume crops in a wave of destruction.

In short, bud rot or botrytis is the most common fungus found inside indoor gardens. The grey mould flourishes in moist temperatures associated with most grow rooms. The damage this grey mould can create is compounded by humid climates, which typically are grow room environments that have humidity levels above 55%, day or night!

Botrytis starts inside the flower and is a little difficult to spot at its onset, however, if you have seen it once you will never forget it! It starts out hair-like, similar to the fluff you get on your laundry and on your clothes. It is primarily grey, but is also white, blue and green in colour. As this fungi increases and grows and consumes, it turns the fruit and flowers, or the areas affected, into a slime. In less humid environments, the affected areas turn dark brown and die back, with a dusting of greyish spores. In these cases the bud rot is in fact almost dry to the touch and breaks up when touched, releasing more active spores. The spores of botrytis are actually present everywhere and attack many plant types and species. Creating an environment in which they can flourish is an extremely dangerous thing to do. It is now widely known that botrytis can also be found attacking stems, leaves, seeds and can even decompose dried stored fruit or flowers.

If botrytis is rife in your grow room and left unchecked, it can decimate your entire crop within 7 days! Where botrytis takes hold of your stems and not your buds which, by the way, is less common in an indoor environment, the stems firstly turn yellow, the outbreak then causes the growth above the infection to wilt and in some cases, causes the stems to topple over completely and die off.

Botrytis is transported via air and contaminated tools. If left unchecked, it can reproduce and consume your crop exponentially. If your grow room is kept below 70°F and your humidity is high – above 55%, then you have created the exact environment that this stuff loves to breed in. If you discover an outbreak, remove all dead leaves and stems as botrytis can often be found harbouring on this dead rotting foliage. Use sterilised pruners to remove infected areas and remove up to one inch more around and below the infected area to ensure that you have cut all of the outbreak from the plant. When removing the fruit or flowers, it is imperative to do it gently and not to knock any other uninfected fruits or flowers with the infected bud, as anything this infected bud touches will in due course also become contaminated. Remove the diseased bud from the garden, isolate it and destroy it. After removing infected areas, sterilise equipment and hands before re-entering the grow room, then increase the temperature in your grow room to around 80°F and make sure that you lower your humidity to below 50%. And if all the above fails, then harvest your crop early and store in a completely dark room as this fungi needs light to survive and to procreate.

Blight

This is a generic term used to describe many plant diseases; fungi typically cause blight more than often a few weeks before harvest. Blight can be identified as, and including, dark spots on foliage, a sudden reduction in growth, yellowing followed by wilting and then death. Much blight can spread quickly through large areas of plants.

Example of Blight

Prevention

Again, as stated before, cleanliness and proper ventilation coupled with proper nutrient balance and no over fertilisation are the key to its prevention.

Downy Mildew

Downy mildew otherwise known as false mildew, affects the vegetative and flowering cycle of a plant's life. It firstly appears as yellow spots on the top of leaves followed by pale patches. Opposite the pale patches on the underside of the leaf is the mycelium and is grey and white in appearance. Downy mildew can procreate in the right conditions exponentially, causing the plant to suffer from a lack of vigour, growth slows and leaves yellow, die back and drop off. This disease is maintained in the plant's system and then grows outwards. Typically, if a plant gets this disease in the early stages of its growth, it is fatal. If you get an outbreak, remove the entire affected plant and destroy it before it passes the disease on to other healthy plants.

Example of Downy Mildew

Prevention

Again, conditions as described above are a breeding ground for this fungi, so keep the grow room and your equipment clean and well ventilated making sure that the humidity does not get too high. Prevention again before cure. Some sprays are available but a lot are ineffective.

Damping Off

Damping off is an actual fungicidal condition also know as pythium wilt. Damping off prevents newly developed seedlings or cuttings from developing. It simply attacks the stem of the new seedling or cutting near to where the stem is rooted into the medium making the stem rot, wither and collapse. It is actually caused by differing fungal species – botrytis, pythium and fusarium. Once it gets started, this condition always results in death. In the early stages, the stems lose their girth, narrowing at the medium line, then weaken, go dark in colour then the circulation is completely cut off, and the end result is kick the bucket time for your seedlings or cuttings.

Damping off is typically caused by fungi already being present in the medium or in fact, the propagator or grow room. Over watering your young seedlings or cuttings and not allowing the medium to dry at all by keeping the medium constantly soaking wet, coupled with excessive humidity, will create the perfect environment for these fungi to take hold.

Example of Damping Off

Prevention

This condition can be avoided by controlling the moisture of the medium. Simply put, over watering is the greatest cause of damping off. Therefore, not over watering is the key to its prevention. It is also advised that you should maintain the temperature for your seedlings or cuttings between 70-80°F. It is also recommended that you use a bright light, like a high frequency fluorescent or even an HID instead of normal fluorescent lights, as bright lights inhibit damping off. It is also suggested to keep nutrient levels to a minimum during the first couple of weeks of growth. Again, good judgement would dictate that you germinate or take cuttings in a clean and sterile medium, making sure that any equipment used is also clean and sterile.

Foliar Fungi Spots

Fungi spots and blotches can affect the stems as well as the leaves causing a mixture of grey, white, brown, black and yellow spotting and blotching. The leaves and the stems discolour and develop these blotches or spots which in turn affect the flow of fluids and many more manufacturing processes, restricting the ability to develop. The overall growth on the plant slows. If your plants develop this at the later stages of harvest, then the finishing time of your plant can be greatly affected, sometimes putting weeks onto your finishing times. If the plant develops this and it takes hold, then it can result in the death of your plants. Again, similar to blight, leaf spot is a name applied to many fungal diseases. This disease can be the result of bacterium, nematodes and fungi. During the growth of the plant, the spotting or blotching can develop differing colours as the plant grows and matures. Leaf spots are often the by-product of damage caused by spraying very cold water directly onto plants that are already hot and well lit under high wattage high intensity grow lights. The temperature stresses related to doing this on your plant can lead to the development of this disease.

Example of Foliar Fungi Spots

Prevention

Once again, to minimise the threat of this disease simply use good sense. Cleanliness as with all fungi related diseases is paramount throughout the whole process. Again, make sure that your entire medium is new and sterile, that if you are reusing items such as your system or propagators, then again it is very good practice to sterilise before use. If you are into foliar feeding or have to nuke your plants due to pest problems, then ensure your lights are off, and have been for a good period of time. This should be for at least 2-4 hours so that the grow room is not too hot before you commence. Also, allow the water that you are going to use on your plants to adjust to room temperature before application. Verify that the humidity in the grow room does not exceed 55%, day or night. If you have a big day and night time swing in temperature, then employ a heater so that you have a maximum of a 10° swing between the hottest temperature and the lowest temperature. Ensure that your plants are spaced well apart and if the problem arises then do not let the plants become overcrowded. Remove any damaged and affected leaves and dispose of them in isolation. Avoid possible over fertilisation.

Algae

This stuff is the slimy green growth that can be found sometimes on the very top surface of your growing medium. Algae needs light and nutrients to grow. It is normally caused as a result of the medium being too wet and coupled with the medium being exposed to too much light. Algae, however, does not pose a threat to your plants or your grow medium, it is simply very unsightly and most growers want rid of it because they do not want to see it or handle it. Algae, although not causing a direct threat to your plants, its presence does attract fungus gnats and other unruly pests

Example of Algae

that can damage the root systems of your plants, and once roots are damaged then disease can take hold with much ease and speed.

Prevention

Once again use your judgement. Do not over water your medium and if you have a lot of very exposed medium, then look to block the light out by covering the exposed areas with a light-tight plastic, such as black and white sheeting. You can spray the medium with a diluted H_2O_2 solution first, which will kill on contact most of the algae, then cover, so light cannot reach the surface or the algae. Without light, algae cannot survive.

Mildew

Powdery mildew affects only the upper surface of the plant's foliage. The first signs of infections are small spotting on the top of the leaves; these spots then develop a grey and white powdery coating that then becomes visible on the leaves, stems and even the new growing shoots. The growth of the plants slows and the leaves yellow, as the disease progresses, the plants die. Plants can be affected

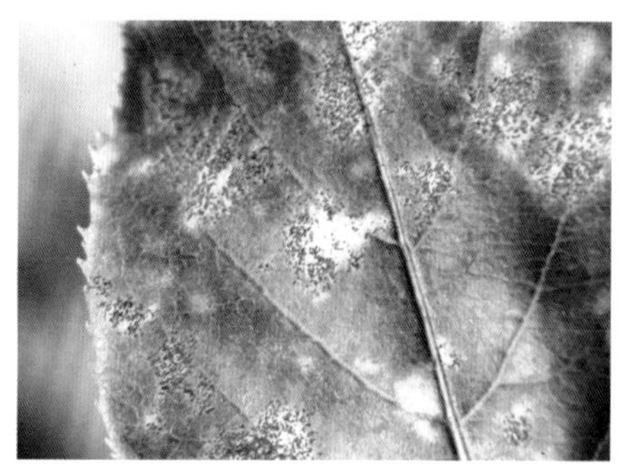

Example of Mildew

for weeks before they show any outward signs of problems. Mildew can also be encouraged if the rootball is dry and the foliage wet.

Prevention

Again prudence is advised; keep it all clean and sterile. Avoid damp, cool and humid conditions inside your grow room. Avoid low light levels and make sure that the root system is well ventilated day and night, as the opposite to this will encourage the problem. Make sure your plants are well spaced and if this disease becomes a problem then remove and destroy the affected leaves. If plants are overcrowded and the symptoms appear, then thin the plants down to allow space for the leaves to breathe. Remove only the foliage that is more than 40% affected and again, avoid over fertilisation

Fusarium Wilt

First signs of this disease are small spots that appear on the older more mature leaves. Leaf chlorosis quickly appears, the leaf tips curl and wilting occurs; after wilting, the leaves quickly dry to a crisp. This disease can affect a portion of the plant or the entire plant. The process can accelerate so quickly that yellow dead leaves dangle from branches. Again, this disease is held in the plant's xylem which is the base of the fluid transport

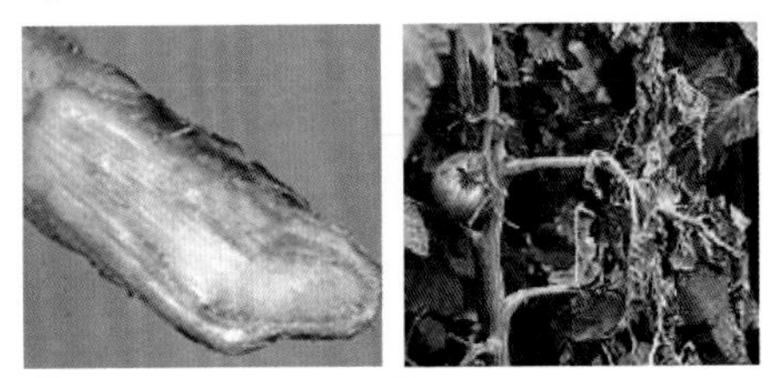

Examples of Fusarium Wilt

system of your plants. The leaves on the plant wilt as the fungi blocks the fluid flow in the plant tissue. So, cut the main stem in two and if you have got brown in the stem, then your plants have it.

Prevention

Be responsible and keep it clean and sterile, only using new sterile growing media. Avoid over fertilisation. Employ the use of Rhizotonic and Trichoderma to strengthen the plant's immune system.

Verticillium Wilt

This is a mysterious disease and sometimes growers never realise that their plants actually have it, as it is not at all obvious, especially if you only check on your plants in the morning. The plants generally wilt progressively during the day cycle, starting off completely fine, then as the day progresses so does the wilt until at the end of the day cycle, the plant has completely wilted. However, during the night period they make a complete recovery and get ready to start the whole procedure again. Lower leaves on a plant suffering with verticillium wilt develop chlorotic yellowing on the margins and also between the veins of the leaf and as this disease progresses, turns them brown.

Example of Verticillium Wilt

Prevention

Once more to avoid this situation use your judgement. Keep it all clean and sterile, make sure the roots are properly aerated and do not over fertilise your plants. It is advised to use products called Trichoderma and Rhizotonic and these should strengthen the plant's immunity to this fungi attack. If you are in doubt that your plants have this disease, then cut the main stem in two and look to see if your plant has brownish xylem tissue and if so, your plant has it.

Pythium or Root Rot

If you get this problem and you are using a re-circulating hydroponic system as most growers in the United Kingdom are, then prepare to weep. Pythium is fungi that causes the roots to turn from white to brown and then into a grey brown mush. The roots at its onset, start to go light brown then darken as the disease progresses. Leaf chlorosis soon follows, then the disease progresses and the older leaves wilt, which is soon followed by the entire plant. During this whole process, growth slows and in severe cases, the root rot progresses up to the base of the plant, turning the stem dark brown. In aggressive cases, and if left unchecked, then these fungi will migrate throughout the entire reservoir and infect every plant that is using it. Entire grow rooms have been completely wiped out because of this fungi. I have seen grown men cry when their plants contract this disease, fully knowing that

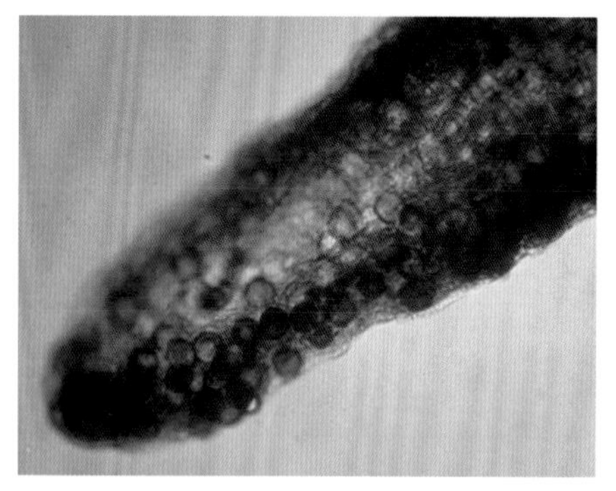

Example of Pythium

they might lose the lot or in the best case scenario get a restricted yield with a load of extra elbow grease to boot. Root rot is common when roots are not sufficiently aerated and therefore are deprived of oxygen.

Prevention

Once again, ensure that you clean and disinfect everything before use and re-use. Employ new sterile growing medium, do not over fertilise, keep pH levels to around 6–6.5 depending on the medium that you are using, and isolate affected hydroponic systems i.e. do not transfer water from one reservoir to another if you are using multiple systems. The disease is passed very quickly through water and an infected reservoir can also infect non-infected reservoirs if you are not clean, dry and tidy. A good insurance policy product to employ is called Trichoderma powder. You apply this to your young plants and in so doing, you expose them to a weak or almost friendly bacteria. Similar to humans having a flu' jab, the young plant's immune system is strengthened through the process, making fungal root outbreaks less of an issue. If, however, your plants have managed to contract this disease, then the best method of controlling this fungi outbreak is to invest in an UV filtration system (see ozone generators section) and if your plants are suffering with this problem, go to this section now!

Chapter 11
Cuttings and Clones

Many advantages are apparent when choosing to take cuttings to generate new plant stock. The main benefits are that the clones taken from the plant are almost exactly that – clones, i.e. the cutting that you take from the mother plant will have almost identical characteristics, so quality, size, yield, taste, sex, strength and vigour to name a few traits are passed down to the clones. Therefore, all the important factors are already known as they follow the attributes that the mother carries. As the characteristics are already known, the cuttings will also have these same traits and grow to a similar shape, size, strength, so on and so forth. So in effect, the space in your grow room can be more efficiently utilised. The other very big benefit is that no further cost apart from sundry equipment is needed, meaning that you do not need to make another outlay on purchasing the seeds, as new plants can be generated from the existing stock of plants. Many new plants can be grown from your original stock for an absolute minimal cost.

Any part of the plant can be used as a clone, as long as they encompass a growing tip. However, some parts make for better or more pertinently, easier rooting clones than others. It is widely believed that the main centre head of the plants and the very top arm tips make for easier rooting cuttings and therefore better cuttings. This is rightly believed due to the fact that these parts of the plants have a higher concentration of auxins (which are growth hormones) compared to the lower canopy of the plants, and therefore, make for better clones that are more likely to root. Now, although the science here is correct, the science of nature is a completely different ball game. Through personal experience, although the tops of the plant are very easily rooted, the lower canopy side branches have more vigour once rooted than their counterparts from the top.

The reasons for this may be many, but an obvious one is that the lower canopy compared to the upper canopy, has had it hard. The lower canopy is struggling for existence, therefore is working hard to survive compared to the tops, which have got it very easy – completely pampered is somewhat of an understatement. So, when you take cuttings from the lower canopy, these cuttings have within them the motivation gained via struggle, and when given an opportunity to become fully fledged plants themselves, they literally fly out of the starting blocks and continue to exhibit this inherent motivation throughout their whole existence. When compared to clones taken from the tops of the plant, which at first make a better cutting, they have however, inherited a type of laziness and lack of motivation that the lower canopy cuttings do not possess. So with all the above in mind, if you could extrapolate this over the course of a few generations, you would obviously only be taking the cuttings from the best of the stock plants. Then in turn, the very best of those cuttings would become the new stock plants, and through taking cuttings from the cream of those stock plants, you would, before you know it, 2-3 generations later, have the best of the best of the best, all with turbocharged motivation and momentum for their survival. The end result would be supercharged stock plants that will quite frankly be the doggies' nuts!, and truly professional specimens that any grower would sell their mother-in-law for! ☺

Taking a Clone

So, if you are conservative and you want to play it safe, then take the cuttings from the tops, but if you are like myself and want to be a bit more radical and discover something different, then take the cuttings from all over the plants, especially the lower canopy. The clones should be taken from the softwood stems; avoid the older, harder, woodier stems or branches. The term softwood applies to the soft younger green stems and shoots. These softwood stems are easiest to establish and root.

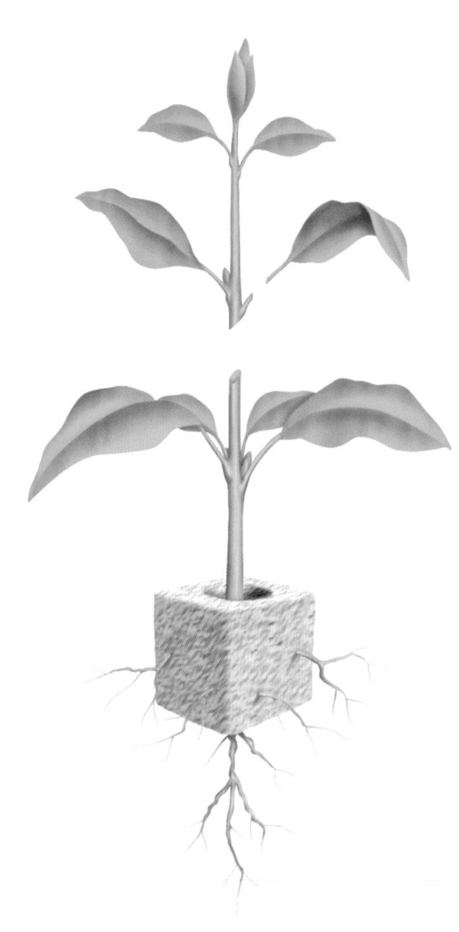

As the plants and the stems develop and mature, they then become known as semi-ripe and once fully matured are known as hardwood. It is important to note that when taking clones, you only choose the healthiest specimens and that they have at least 3 sets of leaf nodes. Any smaller than this will take, root and make a cutting, however, you will find that these will be slower to root and establish compared to their bigger counterparts.

With this in mind, this would also be the case with too big a stem cutting. If you take a cutting that is bigger than this, then the mass of the plant has difficulty nourishing itself through its limited root system and causes great stress upon the cuttings and sometimes results in complete failure. So with most plants, the perfect length is between 5-7 cm in height.

When making the cut from the plant, make sure that your hands are clean and if you roll your own and have handled tobacco, then sterilise your hands or use surgical gloves. This is to avoid any possible transference of tobacco mosaic virus which is rife in fresh or dry tobacco. This virus will quite simply bugger up your newly taken cuttings. On that subject, a story springs to mind of one grower who was completely dependant on others for his supply of cuttings, which was an obvious source of great irritation to him, as the standard of cuttings obtained from others were definitely lacking in quality and vigour. This grower had years of growing experience under his belt, however, for the life of him, although he tried almost every crop, the cuttings were complete failures and the success rate was pathetic; what did root was not worth breathing on. He had meticulously isolated any and every possible problem that could be causing the lack of success, but still no matter how hard he tried, he could not get clones to take. After many conversations with him, it materialised that, lo and behold, he was a

great fan of making his own cigarettes. He was using and handling tobacco even as he took the cuttings. So I quickly pointed out to him that tobacco carries a very dangerous virus known as TMV or tobacco mosaic virus and that if you handled tobacco regularly and then handled your plants, the virus can jump from the tobacco to your hands, to the plants. This virus, if applied in strong concentrations to young plants, will restrict, suffocate and generally slow the plants' efforts to survive. The plant is diverting the majority of its available energy to fight against the threat of death that the virus has created. The end result is either very weak, sick cuttings, cuttings that won't root at all, or even dead cuttings. Or cuttings that take weeks on end to do anything at all. Once the error of his ways was pointed out to him, he sterilised everything related to his grow room and all equipment that was used for taking the clones; he sterilised his hands before proceeding and was back in my shop some two weeks later on his hands and knees in an act of mock worship, I kid you not! ☺ This chap took years to discover his mistake so you don't have to. A true martyr and now never forgotten.

So use your common sense and keep everything clean and sterile including your hands, and do not smoke anywhere near your plants or your cuttings. A sterile disposable scalpel makes a great tool for taking cuttings, and once you have taken your cuttings, then dispose of the scalpel safely, use a brand new one if you want to take more at a later date.

Step 1

Now when taking a cutting, firstly, use clean and sterilised snips or scissors, go around your plant and effectively give your plants a big haircut and remove from the plants a number of stems, or giant cuttings. These cuts off the plant should be at least one internode longer than the desired length of the cutting. Stick this larger and longer cutting straight into a glass of fresh, clean water. It is a good idea to take as many of these cuts as you can get into the glass, then take them to a sterilised cutting board and lay them out one at a time, individually. Then, using your sterilised scalpel, cut the third set of leaves off at the point it connects to the stem, then after removing these lower fan leaves make a cut at a 45° angle on the third internode; when I say on the internode I mean cut a 45° angle directly though the internode, start above it and end below it but cut through it. The reason for this is that the internode has a lot of growth hormones or auxins relating to the fact that the stems for the leaf grew from them, so this junction has a lot of excess energy at its disposal once you have removed the leaves from it. If the next pair of leaves from the bottom are big, simply cut the leaves (both of them) in half to reduce the amount of stress that these big leaves would put on the stem to maintain their growth i.e. the more leaves on the plant, the more stress these leaves levy on the stem to ensure their existence.

However, you do not want to strip the cutting completely down because if you leave little or no leaves apart form the top centre set, then the plant will not be able to photosynthesise at the rate it needs to establish roots quickly for its development. So you want to ideally leave the two centre new growth leaves at the very top then also, if the leaves are not overly large, the next set should also be left. However, if this set is overly large, then simply cut both leaves completely in half from side to side. In effect, you are reducing their size by half; the half leaf that is left will function in the same way as the whole leaf, but will reduce the stress ladened on the stem for its survival.

Step 2

Once you have done the above, dip the bottom of the cutting into a rooting hormone and also, if possible, some Trichoderma solution to reduce any possibility of fungi attack. It is advisable to use a liquid or gel rooting hormone over a powdered one, as powders tend to clog up the cells and prohibit the stems from breathing properly. Once dipped into these two solutions, push the end into your already prepared growing medium. It is also advisable to administer with a pipette, a drop or two of the rooting hormone into the hole inside the rockwool cube. This ensures that the very end of the cutting has contact with the rooting hormone. One inch rockwool cubes are a very popular medium to use; they are sterile and make for the perfect environment for your cutting to root into. If you are going to use these cubes, then you should get them ready long before you even make the first cut to your plants. For preparation, when using rockwool or Jiffy-7 peat pellets, follow the instructions as thoroughly explained in the propagation transcript section at the back of the book or Chapter 3 Propagation. It is worth noting that you should use the rooting hormone very sparingly – as you guessed it – less is best and a lot more effective. In fact, if you take anything away from this book relating to hydroponics it is just that – "less is best and a lot more effective". These are very wise words indeed, so remember them when in doubt. It is also worth noting that as soon as you make the 45° angle cut through the internode, that you dip it the moment after making this cut to restrict any possible embolisms caused by the cutting process. An embolism is when a small bubble of air is sucked up through the stem, effectively preventing the clone from drawing water and nutrients up the stem. So be swift, from 45° cut, to dip, to placing it into its new medium. The use of a rooting hormone and Trichoderma protect the cuttings from potential fungal diseases, seals the cut tissue, protects the initial root tissues, promotes the root cell initiation, and also helps feed the young roots.

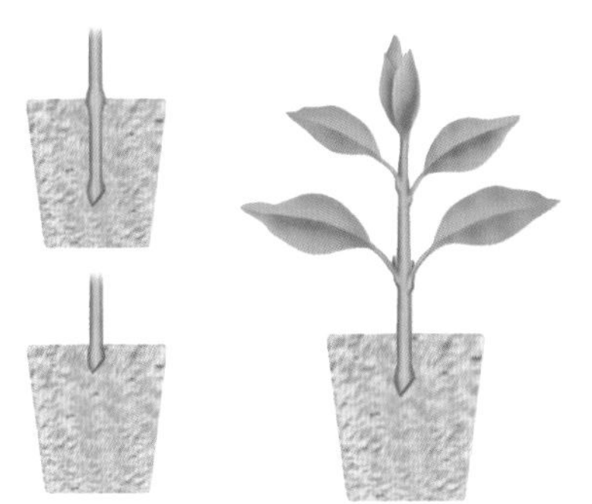

The freshly cut stem should be pushed approximately 1-2 cm into the growing medium. Once you have completed this, a callous, which is a basal swelling, will form; it is completely natural and is the way the plant heals its wound readying itself for the development of its own roots.

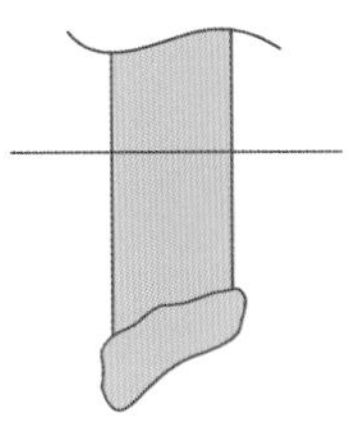

Once the callous has formed, the roots grow from this point soon after. If you have very close internodal growth on your cutting, then you need to strip maybe two sets of internodal growth off the stem of your cutting to get them to be 5-7 cm in height. It is advised that you make sure that any strip internodal growth is beneath the surface of your growing medium, as airborne infection can be created under the surface level, but also that it presents another internodal site that the roots can emerge from, as well as from the 45° angle cut. It is also worth noting that when pushing your cutting into the new medium, that the hole made is big enough to take the cutting. This will prevent the loss of the important rooting hormone gel and other solutions that you have used. If the hole is too tight, these solutions are very easily rubbed off.

Step 3

After you have completed the above, place the cutting and the new growing medium into a propagator with the lid on and the air vent shut. The first four days of propagation are the most critical and this is the time that if they still show signs of wilting, then you have got problems – either an embolism or if it affects many cuttings, then an environmental problem, like a fungi attack.

The first day, the cutting will wilt and look a little sorry for itself, but the leaves remain green and healthy, just wilted.

The second day, the cuttings should by now show signs of improvement and although the larger leaves may still look wilted, the top of the plant however, should show signs of moving up towards the light.

By the third day, the cuttings should be standing upright to attention with no visible signs of wilt and the leaves are quite able to support themselves.

On the fourth day, the cuttings should look at their healthiest with leaves stretching out and up, erect and arching upwards towards the light; the leaves are green and healthy. If this is not the case, then make sure that the growing medium is not too wet or for that matter, too dry and that the humidity is around 70-90% and confirm that the temperature is around 24ºF.

After checking the above and after another 4-6 days, if the cutting is still looking unwell and shows signs of wilting, then carefully check the growing medium for any roots. If your cutting shows no signs of any growing roots then you will need to start again. It is advised to disinfect everything that you have used and that you use a fresh bottle of rooting hormone and a new batch of Trichoderma.

During day 4 to day 7, little above ground level really happens apart from the wilt and recovery of the leaves, however, below the surface, the division of the cells at the base of the cut will have initiated the callous basal swelling from which the roots develop.

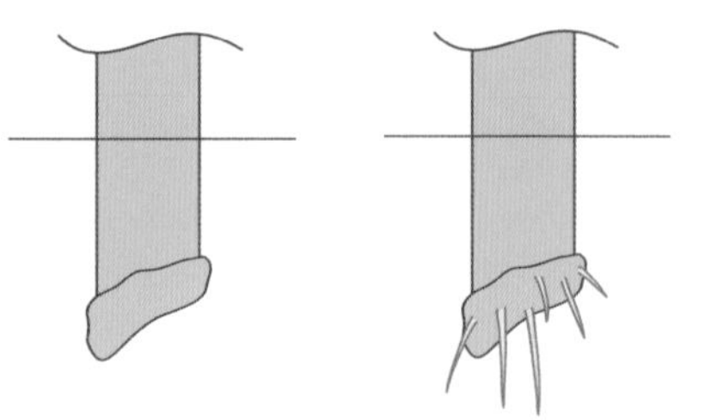

During week 2, the cutting should have developed enough root for the new plant to develop new growth above ground level spurred by its new independence and ability to feed from its growth of the newly formed roots.

During week 3, the roots should be protruding in abundance from the rooting medium and overtly fresh growth will be present. It is now time to root

into a bigger substrate. See propagation transcript or Chapter 3 Propagation.

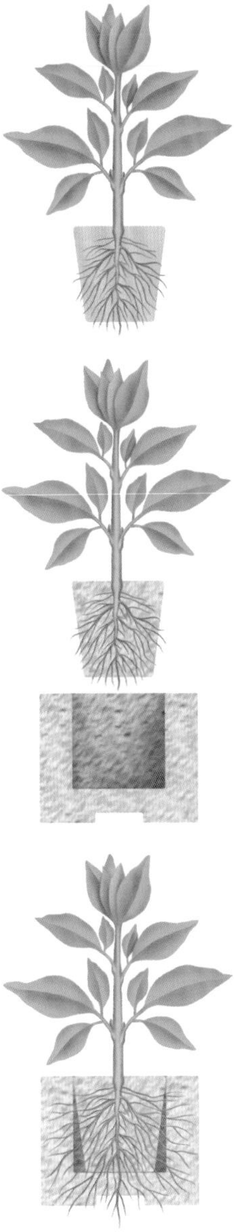

Mother and Stock Plants

The taking of clones requires firstly the existence of either a mother plant or stock plants. As the clones will show very similar characteristics to the mother or stock plants, it is important to select the best specimens to clone from. It is also important that the clones should ideally come from a mature mother – one that has developed a good-sized rootball.

When taking clones from a mother plant, never strip more than 40% of the plant's overall mass. If you take more cuttings than this, the mother can go into acute shock and the growth of the plant will temporarily be stunted, plus the shock can induce all sorts of differing behaviours from the mother. It is also worth noting that a mother should be given time to recuperate before you take any further cuttings from the plant. To get more cuttings per plant, it is a good technique that 2-3 weeks before taking your cuttings, to remove all centre growing tips. Once you have done this to your mother plant, the stem will branch into two new growing stems and shoots, and if you do this again, it will branch off twice again, so one cutting site can actually become more than 4 sites in no time at all. If you do this over the entire plant, then you can get up to 4-10 times more cuttings from the same plant. You must however, let the mother regenerate and get over any shock before harvesting the new stems for cuttings.

Alternative Method of Cloning - Air Layering

This technique is a more time consuming and arduous alternative method of rooting a cutting. It can, however, be beneficial as a means of reducing the height of any stretched plants. Either remove

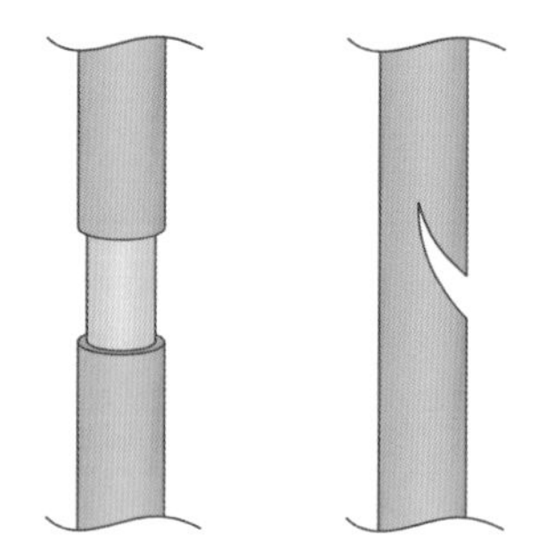

approximately a 1.5 cm segment from the outer layer of the stem or insert an elongated cut near the top of the plant or the top of the side shoots.

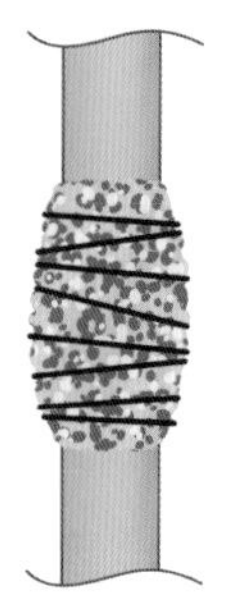

Bandage the incision with vermiculite, perlite or rockwool and secure with muslin fabric and bind with thread.

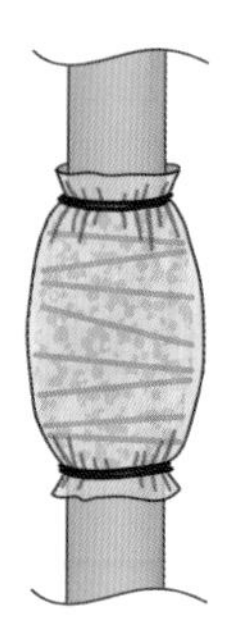

Insulate the entire bandage with clear plastic like cling film, and bind at both ends to ensure humidity remains high around the incision.

In a few weeks, roots will have emerged and will be visible through the plastic. Carefully remove the plastic and with a sterilised scalpel or scissors, cut just below the emerging root system.

Further propagation is advisable for your new cutting.

Chapter 12

Pruning and Training

Depending on the environmental factors of your grow room, some growers find it a necessity to manipulate the natural size and shape of their indoor cultivated plants.

For example, if the grow room does not have sufficient headroom to allow the plants to grow at their natural rate, size and height, a grower would be forced to employ pruning techniques to keep the plant from outgrowing the height of the space provided. Likewise, the same principles would apply regarding the width of the space available. The end result is to cut back or prune your plants to a more manageable size and shape.

A lot of growers cannot get their heads around performing this procedure; after all, these plants are their little offspring that they have so lovingly nurtured, so to cut them back in their eyes is almost criminal and an act of sin. But the truth is that if you let the plants outgrow your lights for example, the plants would have been far better off, and would have produced a much bigger and better yield with less stress put on them if you had cut them back when the time was right.

This nipping out and cutting back, or even pruning is very common in nature. Plants, especially young plants, are being constantly gnawed at by all sorts of animals and pests, so the plants' genetics are quite used to having the top and even the sides chomped off! The plants will recover but moreover, will develop better if your space dictates such an action.

As mentioned earlier, there is a hormone called auxin and it is present in the growing shoot and internodes of plants. It is very concentrated in the main centre stem and this hormone acts as a growth inhibitor preventing the growth of the side branches. These force the plants to grow upright via a centre stem then the branching off of the centre stem occurs. Now if you remove the main centre shoot of the next two branches that sprout from the top, they will then become the inhibitors, so remove the centre growing tip, otherwise known as pinching out. This forces the plant to develop its branching out earlier, thus providing the grower with shorter, stockier and bushier plants.

Pinching Out

To pinch a plant out you must remove the growing tip below the internode point. This effectively decapitates the top and centre growing shoots which then forces the plant to develop two branching centre growing tips instead of one. If again, later on you remove from these two branches, the centre growing tips, then the plant is forced to develop two more centre branches per

original branch, so in effect you will then get a plant with four centre growing branches and tips.

Each time you pinch the plant out, the more branching will occur, restricting the overall height of the plant.

Many growers pinch out the entire plant to encourage overall branching and new stem development. If you are looking to keep a mother plant, it is recommended that you regularly pinch out all the plant's growing tips and this ensures that you have many new stems to take cuttings from.

Pinching out is used to restrict the height of the plant; it is also used to make the plant thicken out and become stockier and bushier.

It is recommended that if you know that your plant has a tall genome type, that you should pinch the plant out at least twice before flowering. If your plants have bolted and stretched due to the light being too far away or the temperature is too high, then again it is highly recommended that you pinch the plant out continuously or moreover, to actually cut your plant down in size by approximately one third.

Removing everything above a certain height and then pinching out any branches that have not already been cut back will then turn your stretched and stressed lanky plant into a cup winner. So do not think twice – this is what you have to do to turn your stretched and elongated plant into something that is worth growing.

Stretched Plants

One word of comfort – if you have stretched your plants then don't panic. Some seasoned growers actually create an environment to stretch their plants on purpose after establishing the plants. They then go to town on them, effectively cutting them almost in half in terms of size and shape; they then change the environmental factors to suit perfect conditions and carry on. The end results can be quite stunning, but the only reasons that they are, is that they cut the plants back hard, then carry on growing them correctly.

If you do not cut your plants back for whatever ethical or unethical reason and they have stretched, then you simply know nothing about growing plants indoors. And believe me I have met many a grower with this short-sighted mentality. This is the answer – employ it if your plants and environment dictate it.

On this note, you can get away with pinching and cutting back a plant up until the first to second week of flowering. After this, you are advised not to pinch out or cut back; what you now have to do to curb the height is to train them.

Training Plants

Training a plant is again quite common, especially if your plants are growing taller every day, and that you have gone past the point of pinching out or cutting back due to already establishing fruit or flowers on your plant. If this is the case and you are running out of height space between the plant and the lights, it is a necessity not to allow the plants to grow higher than the light source, which means you have to force them down and train them back.

This is done by gently and gradually bending the plant's centre branch and branches back towards the ground. Growing plants are quite pliable and as long as you are not brutal with them, they will accommodate a 30-45° bend without snapping. Simply bend your plants from the growing tops to a point where, if you were to bend the stem down

any further, it would break.

At this point, tie the plant back to canes or a weight on the floor to prevent it from bending back to the upright position. After a couple of days, the plant will have adjusted to this new growing position and once more you can force the stem back some more without breaking it, tying it back until it has got used to this new position. In doing this to all stems above a certain height, you can restrict the height of your plant, encouraging it to grow sideways and down on itself and then up. Some growers employ this technique purposefully to gain better control of their plants. This is typically because of the confines of their particular grow room.

This technique of training and bending plants is again only recommended if you have a tall genome species of plant, or that you have gone too far into bloom, therefore cannot pinch out, or that your plants are simply outgrowing your lights. You should start the process of bending the plant stems before it is a necessity to do so, i.e. if you can see it coming, then act now and not when it is too late. The same theory applies to pinching out and cutting back.

If when reading this book you suddenly realise with a sinking feeling in your gut that this is your current predicament, take a deep breath, because the only choice you now have, is a radical one. If your plants have outgrown the lights and are already above them with good sized fruit or flower formations, then you have to employ a technique of snapping!

If correctly carried out, this technique is not as bad as it sounds. Approximately 1-1.5 feet beneath the grow lights, with your fingers and thumb gently, but persuasively, snap the plant at a 90° right angle, and when doing so, make sure you do not over snap it, and although your plant stem is broken, it is still in one piece. Gently do it in a way where the plant can just about support the 90° bend. If you have over snapped the stem, then employ some type of support to keep the snapped joint at a 90° angle.

When employing this procedure on many plants at the same time, you can make an interconnecting canopy directly under the lights using the plants to support the other plants. After 2 to 4 days, the plants heal and the right angle bend becomes a stiff joint and the plants, although put through some stress during this period, will carry on growing as if nothing has happened but with the greatly added benefit of being bathed in light, which in turn will produce a much bigger yield.

When snapping or tying plants back, due to the weight of the plant now being off centre, it is recommended to cane the plants for extra support.

This support is especially needed if they are heavy with fruit or flowers, because after the bending process, the stem, due to additional weight, can cause a break if your fruits or flowers become too large.

This is also the case with snapping. The stem can break again due to the additional weight of the yielding fruit or flowers.

For this matter, it is good practice to cane any heavy yielding plant so that the plant is not restricted by a particular weight levied on its stems, i.e. the more support it has during heavy flower or fruiting, the less stress put on the plant and the bigger the fruit or flowers will be. Not just that, there is nothing more gutting then seeing a super prize bloom snapping off under its own weight before it has finished flowering. If in doubt, support them.

Chapter 13

Breeding Your Plants

Since mankind took up agriculture and became horticulturally aware, we have been breeding plants, sometimes purposefully, other times not. On the other side of the coin, great Mother Nature has been breeding plants since the beginning of time, and we are the exponential outcome of this experimentation that she naturally performs.

The first original farmers realised that taking seeds from only the best and healthiest or heaviest yielding plants, and not from their opposites, gave better results; this process carried forth over time and many, many generations differentiated plants into separate varieties. These new varieties then took on differing characteristics which then had different uses and could thrive under differing environmental conditions. This is basically how breeding started untold years ago.

In recent more modern times, true scientific breeding began with Gregor Mendels' experiments on plants' inherited characteristics. Gregor Mendels' experiments crossed peas with different characteristics and discovered that the offspring inherited definitive traits from their parentage in a logical, statistical and predictable way.

DNA

Modern science now tells us that each cell contains a set of chemical blueprints referring to every type of aspect of its existence. These blueprints or chemical codes are called chromosomes and consist of long double strands of sugar, based on one of four amino acids. Sets of three of these amino acids form genes which are read by structures in their cell, which in turn directs it in its life processes. These chromosomes are typically found in pairs in most cells.

In most flowers and seed producing plants, half of each pair of chromosomes is contributed by the male through pollen and the other half via the female through the flowers. Flowering plants in general have 10 pairs, which are 20 chromosomes. These chromosome genes are queued in a specific order. Corresponding to this, the other members of the pair has a gene mirrored in the same location. In some cases, a single gene can be responsible for a particular characteristic, and in other cases, several genes are the perpetrators of particular characteristics, typically in a complete series of reactions.

Characteristics

Approximately 2% of the time, genes migrate from one member of the pair of chromosomes to the other. This is an important fact, as it gives individual chromosomes the mechanism for changing the information relating to the characteristics for how they are coded, and should be noted if you are going to undertake breeding. There are many factors when considering plants for breeding, and this would be a greatly simplified task if only one trait or characteristic was involved. However, this is not the case, and characters like yield, potency, maturation time, colour, taste, height, aroma, resistance to disease or pests and vigour, as well as others, go into the mix.

The most important characteristic when breeding is

called partial dominance. Now, a plant that is bred true has most of the corresponding genes on each pair of chromosomes of the 10 pairs, so they have the same information in balance. But plants with parentage from differing varieties which are crossed represent hybrids, and much of the corresponding gene sets of chromosomes have conflicting information.

As an example, a first generation cross which is also known as a F_1 hybrid can contain genes from a parent which may have characteristics of being tall and slim, and this same F_1 hybrid may also contain genes from another parent programmed for stocky and short growth. In this case, the offspring should have an equal balance of both characteristics given by either parent. So, if this plant was bred from these two opposites from true breed plants, than you will get offspring with uniformity reflecting an equal mix of both parents, meaning all the offspring will be neither tall and slim nor short and stocky but an equal combination of both. Consequently, you would end up with all very similar plants that are medium in height and stature.

Now shit hits the fan when two F_1 hybrids are crossed and it does not matter if they are the same F_1 hybrids or completely different F_1 hybrids. The end result is completely different looking offspring as the seeds that are produced when breeding in this way have a complete imbalance of partial dominance, and the end result is some plants are tall and slim, others are short and stocky, others are neither but a mixture of both, while some are tall and stocky and some are short and skinny. You basically get a complete mishmash of all the parentage thrown in together for good measure. This can represent a nightmare in an indoor grow room, as some plants bolt and others do nothing at all, making it very hard to raise and lower lights as well as maintaining correct nutrient and pH levels.

Flavour, aroma, potency and yield also fall into this category of partial dominance. So, when more than one gene is involved, there are exponential numbers of possible combinations which affect the outcome of your breeding efforts. To this effect, characteristics are coded on genes which are either recessive or dominant and thousands of plants must therefore be grown over many generations to find a specimen which meets the criteria set by the breeder.

Most growers do not have the space or patience for such an undertaking, however, it must be said that when breeding F_1s together, some of the offspring show exceptional qualities of the parents that went into it but this is very much a lucky dip. If you are intending to breed your own plants, then the first step is to grow both male and female plants together. If you only intend to breed one or two of your plants, then you need to grow the male in isolated conditions.

Pollination

Once the male has matured, you can harvest the pollen from the pollen sacks and store this in an airtight container. Then, with a small children's paintbrush either straight from the male plant or from the stored container, dip the brush to gather pollen on it, and without any violent movements, simply paint it on the receptive female flowers. You can even fertilise parts or even individual flowers this way, leaving some of the plants to produce seed and other parts to produce unfertilised resin filled flowers.

It is important to remember that millions of these pollen particles are airborne when harvesting them from the male. They can contaminate the entire grow room, and for that matter can also contaminate yourself in fertile pollen. If you do not change your clothes and wash your hands, or make sure that one room cannot possibly contaminate another, then you could accidentally pollinate your entire crop of virgin females. So, be careful and be warned, even sneezing if you are holding a brush which is heavily laden with pollen could seed the entire environment.

Procreate

With all the above said, true breeding in the confines of an indoor grow room is actually impossible to manage unless you have thousands of plants and many years to dedicate to it.

However, breeding mutants can be great fun and sometimes very rewarding but also more importantly, it creates new life with new possibilities and new characteristics, albeit unstable. These offspring will be unstable but over time, one day in the future, all mutants will find balance and hey presto, true breed plants emerge once again to carry on the never-ending cycle of evolution.

Hydroponics
Indoor Horticulture

Chapter 14
Harvesting

This subject is not widely covered by growers or authors and like the storing of wine or whisky, many techniques are praised and practised. The most simple and also the most effective method is a quick manicure of the excess leaf matter, then hang what's left upside down in a dry, dark room.

It is widely believed that once harvested, there is no other process or esoteric method which will increase the flavour, potency or aroma of the plants. This is again a simple myth that is easily dispelled. Although some risk is involved in the very esoteric methods being currently employed and true, you could in fact do more harm than good, I believe that the risk being undertaken, compared with the possible heavenly end result, is a worthy gamble. A gamble, however, only for those that know what they are up to. So, if you are going to experiment, use only a proportion of your crop for this undertaking.

Crazy Ideas

Do not get me wrong here, many crazy American ideas are not worth even contemplating as they represent no benefit to your plants! Absurd ideas like boiling the root of a plant, or exposing the plant to UV light during the curing process, or spraying your plants with a strong sugar spray while they dry. These ideas are quite frankly stupid, and in most cases, extremely detrimental to the curing process. Some popular methods are also used, like using microwave ovens, herb drying boxes in ovens and even freeze-drying. It is not recommended that you employ any of these mad, but inventive techniques.

The simple methods typically are the best as with the whole process of growing your plants to full maturity – less is best and represents more. Again, if something is not broken then why try to fix it. To start with, you can either cut the entire plant at the base and give the whole plant a quick trim removing only the excessive fan leaf foliage, then hang upside down in a room kept at a temperature of not less that 65ºF and not more than 75ºF, making sure that it has a regular air exchange once or twice a day, and ensuring that the humidity does not exceed 55%. It is also important to keep this room in darkness, as the light can damage some of the delicateness of the plant's taste, flavour and potency. When employing this method, the whole drying process will take quite some time as the whole plant has to dry out, however, as with wine, the longer the drying process, the better the end product will be.

It is very important not to rush the drying process, as a quick dry normally represents a very harsh and tasteless end product. This way your plant, depending on the size and yield, could be ready in 7 to 21 days from the original hanging.

Another quick way of drying is to strip down individual branches and stems, cutting them completely off from the plant. Then, give them a quick trim to remove excessive amounts of foliage and after doing this, the branches and stems are hung upside down in a dark room and kept in conditions as mentioned in the previous paragraph. Typically again, depending on the size of the harvest, this technique will ensure that your harvest will be ready in around 4 to 7 days. Although this

is a much quicker way of drying your plants, you will lose a little flavour and sweetness. This is currently the most widely used and most popular method of drying plants.

Now some growers completely strip the plants down, harvesting the fruit or flowers, then individually manicure the flowers and place what's left onto drying racks made out of screens which are similar to what you would dry your wool jumpers on. If the flowers are big enough, hang them upside down using clothes pegs. Using this method, the end product will be ready in double-quick time, however, as the time it takes to dry the fruits or flowers is not long, the end result is a harvest that is definitely lacking in flavour, odour, potency and sweetness. Some growers that do it this way even employ a dehumidifier which any herbalist will tell you will quite literally suck out most of the delicate goodness from your much loved harvest. So do not for any reason employ a dehumidifier to dry your crops as what you will be left with will not be so pukka. The very same applies if you use air conditioning in your drying room; again do not do it but be patient and the reward will be well worth it. If you use the aforementioned racks and small flower method, then your harvest will be ready in about 2 days, however, much of the goodness that you worked so hard to produce will be forever ruined.

Patience

The best drying process is the longest, and dried flowers should retain some moisture and have a pliable and supple nature to them. They should not be brittle, or crumble to the touch. To be sure, take a sample of your harvest during the drying process and experiment with it to see if it is ready or not. A properly dried flower will retain up to 10% moisture which is necessary to maintain the flowers fresh pliable nature but also to ensure good aroma, potency and taste. Dry flowers weigh approximately 15-20% of that of the original harvested wet weight.

Flowers that are dried in a hurry i.e. 2-4 days or less, are being dried in conditions that are too hot or too dry or both, and will produce a third rate end product. A flower that is dried too quickly will be brittle and crumble to the touch when handled and if the drying room is too hot or the humidity too low, the outside of the flower becomes crisp while the inside remains moist. Not a good combination to have as the outer shell of the flower will not allow for the inner moisture to dry out and the end result could get messy over time, with the possibility of bud rot.

Manicure

Before the drying procedure, one has to manicure the plants. This process increases the flavour, aroma and potency of the end product. It is very easy to do and the ideal way is to remove unwanted excessive foliage so as not to dilute the potency of the flowers that have matured. On freshly chopped plants that are hung upside down so as to expose the leaf stalks, remove by cutting away the larger fan leaves at the base of their stalks, then progressively cut back smaller leaves. It is advised to only give these a light trim as during drying, these will curl back into the flower and retain the shape and density of the buds. So do not overdo the trimming as much flavour, taste, aroma and potency are on these smaller leaves. Do not remove any leaves that are part of the flower for the same reasons as above. These smaller leaves that are actually a part of the flower will be sucked back into the bud during the drying process and will maintain the shape and structure of the flower. Some growers, after completing the drying process, give the flowers a final trim to ensure that the end product looks its best. This also maximises

its flavour, taste, aroma and potency.

Most growers use scissors or spring loaded scissors to perform the manicuring of the flowers. These are very effective and very popular but over time can become tiresome. The Dutch have invented motorised scissors that are very easily operated and you can even perform the most delicate manicuring using these electric power scissors or shears. The Dutch have also invented a machine called a clipper which works like a lathe or lawnmower with a blade that rotates at a very high speed where you simply turn the flower near the cutting edge to remove excess foliage. Although these machines for commercial applications are a great time saver, they are very hard to use and you can lose a lot of good flower to them as they really do strip the plant in no time at all.

Flowering

Most flowering plants go through a few stages of the flowering process. Flowers appear at first, and then over time, new flowers develop close to and around the original first wave of flowers. At each leaf node, flowers also form along the main stem and branches. The buds then start to fill out so flowers upon flowers create clusters that are thick with pistils looking for pollen to fertilise them. These pistils are white and translucent and over time begin to change colour, wither and turn from white to red or brown. Now it is very important to be patient as just as a cluster of flowers looks like they have finished and three quarters have changed colour, it is very common for a new wave of flowers to appear and more growth is undertaken by the plant. These new clusters typically concentrate in many of the bare spots that the plant may have. Then, successive waves of flowers can appear and this process can last for weeks after the original finished flowers are formed. As the flowers and clusters of flowers begin to close, then the calyxes start to swell. The calyxes are false seed pods as the flowers have not been fertilised, therefore, no seed can form, however, these calyxes become completely covered with resin glands.

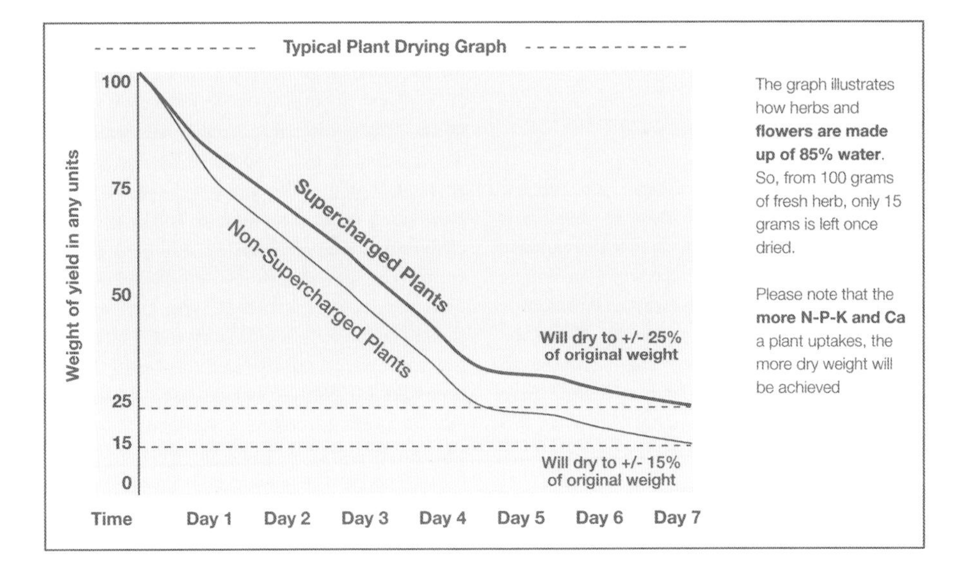

When mature, these glands sparkle in bright light like millions of diamonds. If you were to examine these individual glands under a microscope, they would appear clear, then over time as the glands turn amber, this is the time the buds should be harvested. If you are not employing the use of a microscope then wait until two thirds to three quarters of the flower has changed colour from white and green to red and brown.

It is of the utmost importance that no bud should be harvested before it is ready. It is almost worthy of noting that plants and varieties of plants mature differently and the same applies to the pattern of maturity. To illustrate, some species of plants mature as a complete whole, so you can harvest the entire plant in one fell swoop. Other plants mature, especially under lights, from the top downwards, so in this case you should only harvest the parts that are ready. In so doing, this then allows light to reach parts of the plants it could not reach before, therefore, allowing for more expansion of the undeveloped existing buds. If you get a plant species like this, then you can end up in a situation where you could be harvesting your plants over the course of an entire month.

Esoteric

Now for the more adventurous and esoteric growers that have more flowers than their requirements, I will share a technique that possibly only a handful of growers in the world are willing to undertake and exploit. In fact, this technique is so rare I only know of one person that does it and from first hand experience, this is the most specialised drying and curing procedure I have come across, and to the best of my knowledge, this one grower pioneered it.

This drying and curing technique starts off in the usual fashion where either the entire plants are hung upside down with the excessive foliage removed, or the plants are stripped to stems and branches, with excessive foliage removed. Kept in a room maintained around a temperature of 70ºF and humidity kept around 50%, air exchange is bi-hourly and the plants are hung in a fashion allowing plenty of space for them to breathe. They are kept in complete darkness so they have dried sufficiently that they have still retained moisture, are supple to the touch and would be hard to break apart using your hands alone. So, in effect, they are possibly 80-90% ready or 80-90% dried compared to the conventional methods as explained above. These flowers are then cut from the branches and stems and are placed loosely in an airtight glass jar. These are the same jars that are employed to make preserves and jam; the large jars that have a wide neck with a rubber seal that clips down tight similar to a bottle of Grolsch beer.

The buds are placed inside the jar but not packed tight, but also not loose enough to shake about or rattle. So if you can imagine the flowers used to fill the jar would take up approximately 70-80% of the volume of the space available in the jar, in effect, the jar would be completely full but with at least 20% of air left in it available for the flowers to utilise. This grower also breathes out deeply into the jar before he seals it so the jar has a mixture of air and CO_2 in it. Now here comes the crazy part. However, it is not that crazy, as whisky and wine enthusiasts have been doing something similar for years. This grower goes out into the woods and finds a suitable spot near a large tree or other similar landmark, then digs down with a little hand held spade and literally buries his jar full of flowers approximately 3-4 foot underground. It has to go this far down to ensure it is stored at a very constant temperature similar to wine in a wine cellar. This far underground, no matter if you have a frost or it is sunny, the temperature remains constant. This grower then leaves the jar

underground for as long as he can manage. I can tell you from first hand knowledge that he dug some up that had been kept underground for more than 4 years and the end product had metamorphosised into something almost unearthly. It had retained, compared to other flowers, an exceptionally strong aroma, taste, potency and flavour that was off the Richter scale. To this day, no flowers can compare to what was witnessed that beautiful day. This grower has flowers kept in the same fashion underground, from 3 months old, all the way up to 6 years old; he even has that same bud that I was privileged to see which is now 2 years older. Only time will tell what this might become.

Let's face it, wine and whisky both benefit from this practice – slightly less crude I agree, but nonetheless the principle is the same. So your plants also would greatly benefit – that is if you have the excess and the time to spare! It is my hope now through sharing this technique with you, that many growers will also follow suit, and before long, vintage flowers, like wine and whisky, will be of great value and worth one day in the future.

Hydroponics
Indoor Horticulture

Chapter 15
Equipment

We have already covered the differing hydroponics techniques available, but this leaves the rest of the subject of grow room equipment untouched. In the following chapter, we will cover topics of lighting, light movers, environmental controls, extraction, CO_2 and new products like ozone generators and UV filters.

(HID) High Intensity Discharge Lighting

The best and most popular form of lighting for the indoor garden is high intensity discharge lighting. They simply grow outrageous high energy plants the best. Like for like, high intensity discharge lights outdo all other lamps available to date. They also consume less energy per lumen output compared to any other light so far. So, all in all, they are the brightest and the cheapest to run but also have more usable light spectrums for high energy plants than other lights available.

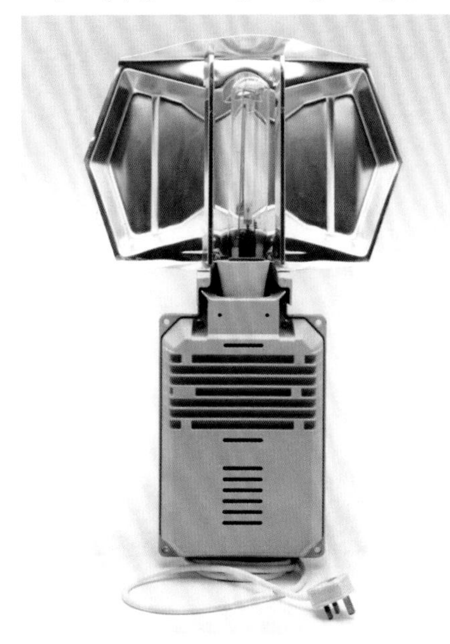

Example of a HID Light

The most widely used HID light is the high pressure sodium, quickly followed by the metal halide. Mercury vapours are also high intensity discharge, however, they are seldom used now and most growers who do use them, now use them for propagation only. Mercury lamps were the first HID lights to emerge and were a marked improvement on their fluorescent predecessors. However, the metal halide quickly outstripped the mercuries, then the nail in the coffin was the use of high pressure sodiums. The mercury vapour is the most inefficient of the HID lights and also the most restrictive of the light spectrums, with only a small spectrum that is actually usable by the plants. Mercury vapour lighting also generates more heat per watt compared to the metal halides or high pressure sodiums.

Metal Halide Lamps

To date, metal halide lights are the most effective and efficient source of artificial white light available to the indoor gardener. Available in 250 watt, 400 watt, 600 watt, 1000 watt and even, but rarely used by the indoor botanist, the 2000 watter. The reason this is the less popular option is the sheer heat output of the 2000 watt light which makes it completely impractical for growing high energy plants. Most growers tend to use the 400 watt or 600 watt metal halides. They are mainly used during the vegetative cycle of a plant's life however, some growers use them to supplement high pressure sodiums during the flowering cycle of a plant's life. Smaller 250 watt metal halides are mainly used for propagation of many cuttings and/or the maintaining of a mother plant or two.

Metal halide lights produce light by arcing electricity through vaporised argon gas, mercury gas, thorium iodide, sodium iodide and scandium iodide; all this within a quartz arc tube. The ends of the arc tube are coated with a heat reflective material to control the temperatures during use. Supporting springs found in the neck and at the end of the tubular cigar light envelope, mounts the quartz arc tube frame in position. A bi-metal shorting switch that closes during lamp usage prevents voltage drop between the main electrode and the starting electrode. In general, lamps are fitted with a resistor that stops the lamp from shattering under extreme temperature stresses. The outer lamp casing, which is normally cigar shaped, acts as a protective jacket for the inner workings. This outer jacket produces a constant environment for the inner workings as well as absorbing ultraviolet radiation. The initial vaporisation takes place in the gap between the main electrode and the starting electrode after a high starting current is applied. When enough ionisation occurs, electricity will arc between the main electrodes. The lamp slowly warms and the metal iodide additives begin to flow into the arc stream. After they warm and form into their proper concentrations within the arc tube, a very bright white/blue light is emitted. A metal halide light can take up to 4 minutes to get to its optimum light efficiency.

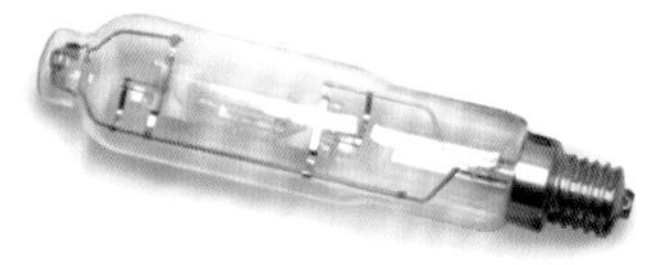

Example of a Metal Halide Lamp

Be warned that if you spray water near any HID lamp, there is a great probability that it will explode. If a metal halide explodes, do not inhale any air, switch off all electricity and leg it with your head down. If a hot lamp breaks, it is imperative that you leave the grow room ASAP as hot metal halide gases will be released into the atmosphere. These gases are a mixture of heavy metal elements, that if absorbed while breathing in, could basically leave you in a perpetual state of madness. After an hour or so, the metal in the atmosphere would have cooled and dropped to the floor and become no longer absorbable through the air. Please note that your plants will not suffer but your health will, so forget getting your plants out of there, just get yourself out ASAP! If a cold lamp breaks, these precautions do not apply.

If the outer shell of the lamp was to crack or break, get rid of the lamp. A crack or a break will allow ultraviolet radiation to leak and this is very bad for you and your plants.

Like halide lamps, if you get a power failure or you switch your lights off, allow at least 20 minutes for the gases to cool before re-igniting the lamp. Never re-ignite a hot lamp as this can stress the ballast and lamp, and you could end up damaging one or both components.

Metal halide lamps have good lumen maintenance and output. However, lumen output as again with all HID lamps, will reduce over time. An average HID lamp should last for at least a couple of years however, over a period of approximately 9 months, a HID lamp, especially a halide, will depreciate in lumen output by as much as 30 percent if in use for approximately 12-18 hours per day. So, although they still work a year on, the brightness of the lamp has decreased considerably. It is in effect a false economy to run an old HID lamp as it is still costing you the same electricity to run it, however, you are now getting approximately 30% less light, which in turn will produce 30% less yield. Change your lamps in your indoor garden at least every 9 months to avoid this non-profitable situation and to maintain maximum yield potential.

High Pressure Sodium

The most popular type of indoor grow light is the high pressure sodium light otherwise known as HPS lighting. They are available in the UK as 250 watt, 400 watt, 600 watt, and 1000 watt options. These types of lights are used primarily for growing high energy plants and are especially good for flowering plants.

Example of a High Pressure Sodium Lamp

High pressure sodium lamps are the most impressive lights when used to cultivate high energy plants. The 250 watt high pressure sodium is mainly used in a closet type of environment where space is at a premium and heat output from the lamp is minimised due to low wattage of the light. The 400 and the 600 watters are the most favoured high pressure sodium lamps. The 1000 watt HPS light is mainly employed if the grower is using a light mover. A lot of indoor growers find it hard to work with 1000 watt lights due to their heat output. This is why they are mainly used on movers to get more coverage with such a powerful light, but also to dissipate the heat as the source is moving up and down and potential heat damage to the plants is minimised. It also covers a great distance, and as most light movers are 1.5 metres long, even when the lamp is at the other end of the rail, the plants are still nourished with dissipated light which is still useable to the high energy plants. For example, if you were to grow high energy plants using a 400 watt high pressure sodium on a light mover, then by the time the lamp is at the far end, the plant furthest away would not be nourished at all by the diminished light source. This is not the case with a 1000 watt light.

High pressure sodium lamps produce an orange/yellow like glow that is similar to that of a setting sun in the summer. Unlike the metal halide, which produces white and blue light, the high pressure sodium is highest in the yellow, orange and red end of the light spectrum. For some years now, and this is still a heavily debated subject, it has been believed that during the vegetative cycle of the plant's life, the high energy plant prefers the blue end spectrum for the growing cycle which is the metal halide. Then, during the flowering cycle, the plants prefer and need the red end of the light spectrum which is of course the high pressure sodium. Although science has proved this to be true, recent findings have shown increased vigour and yield when both lights are used together during both cycles. So although they are adequately nourished using a metal halide for the vegging and sodium for the flowering, it seems the plants are more nourished as both ends of the spectrum are available for its needs. So it would seem that although they do not need sodium/red light during vegetative growth, they can still benefit from it, and likewise with metal halides during flower. The best model that we can study would be the sun. This emits all light spectrums and plants have flourished for millennia as this is the source that nourishes all of life. The conclusion is simple, mix the spectrums; the sun does, so why don't we?

High pressure sodium lamps produce light by passing electricity through a majority of vaporised sodium and mercury and a touch of xenon gas within an arc tube. The xenon gas is used for starting the ignition process. High pressure sodium lamps are completely different from metal halide lamps. It is different physically, it is different electrically, and different in colour spectrum output. A short high voltage pulse that is generated from the electronic starter and the magnetic component of the ballast initiates the starting of the lamp. The electronic pulse vaporises the xenon gas that arcs between two main electrodes. After 3-5 minutes, the lamp warms up and the light emitted is pretty substantial. Like the metal halide lamp, the high pressure sodium has a two bulb construction. An outer protective bulb and an inner arc tube, springs and a frame within the bulb, suspend the arc tube. The outer bulb protects the arc tube from damage and also contains a vacuum, which reduces heat loss from the arc tube.

HID

All HID lights are not in any way like incandescent or fluorescent lights. When they are switched on, they need a period of time to warm up, and to get to full intensity takes approximately 3-10 minutes. However, they also need to cool down before re-igniting them again – approximately 3-10 minutes. Never switch these types of lights on and off in quick succession like you can with incandescent or fluorescent lights.

Once a HID light is on, let it warm up before switching it off and likewise once off, allow to cool down before switching it back on. Simple rules that will prolong the life and output of your lamp.

High pressure sodium lamps have out of all the HID lamps available, the longest life and best lumen output maintenance. After a long period of time the sodium gas bleeds out through the arc tube; this reduces the lumen output considerably. Likewise, the sodium and mercury ratios change causing the voltage within the arc to rise. This over time will create a voltage that the ballast is unable to sustain with the end result being that the lamp will not work.

All HID lamps over a period of approximately 9 months use, will depreciate in lumen output by approximately 20-30%. This is natural and to run an old lamp represents a false economy as you are still paying the same electricity rates to run a lamp which could be 30% dimmer than it was 9 months previously. Reduced lumen output equals reduced yield. For example, if the lamp is 30% less efficient, the yield of the plants will reflect a 30% diminished return. Double the lux and you double the yield, divide the lux and you divide the yield. So, with this in mind, it is best to replace the lamps every 9 months to avoid this situation. Some professional growers are so particular about this that they change their lamps every single crop to maintain maximum lighting conditions. This might be seen by some as a little excessive, however, the growers that do this are very good growers indeed, so who are we to say no need? They think and do otherwise with great results.

High pressure sodium lamps are by far the most widely used HID lamps for horticultural use. They simply grow great high energy plants. In fact, due to their demand in horticultural use, all major manufacturers now produce what is called an agro, planta, or growlux variant of the "tried and trusted" high pressure sodium lamps.

The Son-T agro, the Son-T planta and the growlux variants are indeed still high pressure sodiums but have been specifically manufactured by large companies to be more suited to plant growth. They have basically taken high pressure sodiums and

Typical Lamp Types

1. 1000W Metal Halide
2. 500W Mercury Blend
3. 250W Mercury Blend
4. 600W Planta-T High Pressure Sodium
5. 250W HPI-T Metal Halide
6. 400W Son-T-Agro High Pressure Sodium
7. High Frequency Fluorescent Lamps
8. 600W Nav-T-Super High Pressure Sodium
9. 250W Son-T-Plus High Pressure Sodium
10. 400W Son-T-Plus High Pressure Sodium
11. 400W HPI-T Metal Halide
12. 600W Grow Lux High Pressure Sodium
13. 600W HPI-T Metal Halide
14. 1000W Son-T High Pressure Sodium

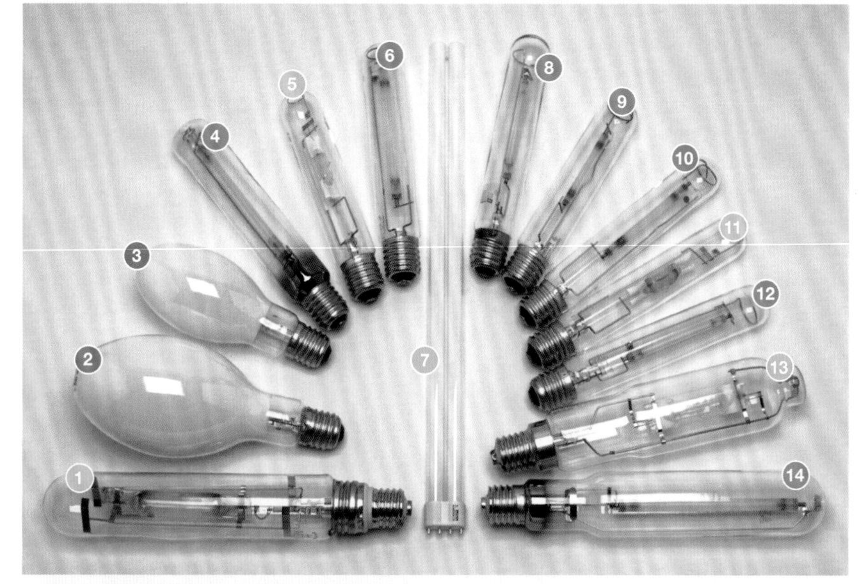

Relative efficiency of some light bulbs

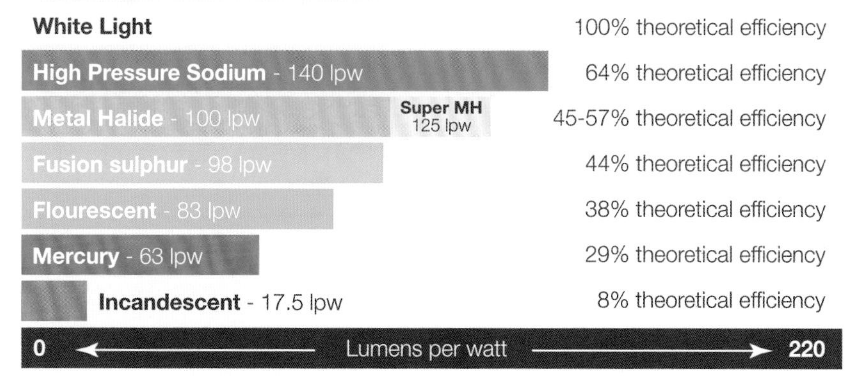

tweaked it to produce more blue end spectrum than regular high pressure sodiums do. So what you have, in effect, is a high pressure sodium lamp which is predominantly red/yellow in spectrum with a dash of blue thrown in for good measure. So you get the best of both worlds. The slight downside of this and it is slight, is the fact that the agro/planta/growlux varieties are not as bright as the regular high pressure sodiums. So, many growers still stick to the regular HPS to maintain maximum lumen output. The really anal growers that are out there and believe me, they are out there, use a metal halide for the vegetative cycle, then switch to an agro during the intermediate stage between vegetative growth and flowering growth, then once flowering has truly been established, they finish the plants with a regular high pressure sodium. This may seem a little extreme but growers that employ this methodology get outrageous results. The other great advantage of this is that because three lamps are in effect being used, the wear and tear on the lamps is minimised. Therefore, you extend the use of the lamps by approximately three times. So long term, this strategy is not that extreme.

Fluorescent Lamps

AKA fluoros, they work by passing electrical current through gaseous vapour under low pressure. Back in the day, fluorescent light was actually the most effective and the most widely used artificial light available for indoor gardening. Some fluorescent lighting has the closest replication of the sun's light in all the spectrums, however and sadly, the lumen output was quite frankly not so pukka! They simply are not bright enough to grow great high energy plants.

Today's budding indoor horticulturist uses fluoros for raising seedlings or establishing cuttings. These lights are great for this job as the colour spectrum is right but also the lumen output is not severe and the heat output is minimal. The result is ideal propagation lights.

There are many different types of fluoros, but these can be broken into two main groups. The low frequency and the high frequency. The high frequencies are now the leading propagation lights on the market. They generate much more lumen output compared to the low frequency and they are also very energy efficient. Low frequency are found almost everywhere; schools, shopping centres, so on and so forth. They are normally two, four, six, or even eight feet in length and are sold in lighting units that have these lamps paired. If you are thinking of using this type of fluorescent unit, you will need a bank of them for propagation i.e. at least three sets of two lamps. And yes, they do work well as long as you have enough of them. The high frequency alternative is mainly used on its own; one complete light over one large

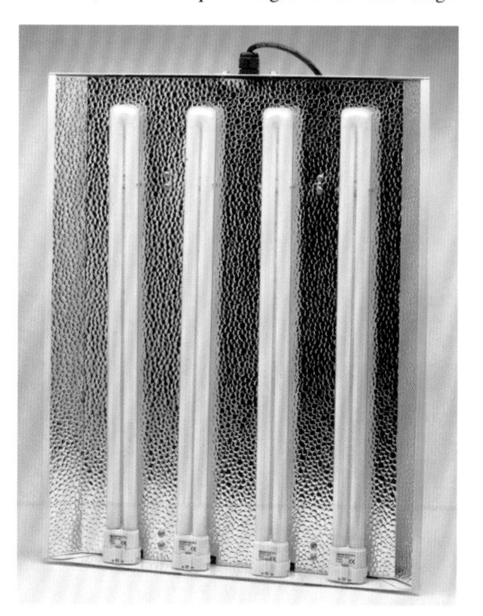

Example of a High Frequency Fluorescent Light

propagator. These high frequency units are also manufactured to take pairs of lamps. They simply are pukka, small, light, and great workhorses. Many growers, once the light is not in propagation mode, use these lights as side or buffer lighting for their grow room, and as these lights are in fact bright enough, report good results through doing this.

In general, they produce the blue end of the light spectrum, however, they are available in many colours; the cool white or blue is the most popular for use with propagation.

These lights, like the high intensity discharge, require a ballast and an ignitor. The ballast, unlike the HID, is small, as is the ignitor. The ballast and the ignitor are typically located within the housing of the unit. So, the lamps, the ballast and the ignitor are built-in, making this unit very practical. The ballast, unlike the HID light, generates the majority of heat, the lamps themselves run very cool, and in most cases the plants can actually touch the flouros with no detrimental effect. The ballast, similar to HID lights, reduces the current in the tube for maximum efficiency. They feed a trickle current to maintain the electricity to the lamp without costing a fortune to maintain. The ignitor, like the HID lights, feeds a pulsed current to ignite the lamps, then a trickle current from the ballast to maintain the light.

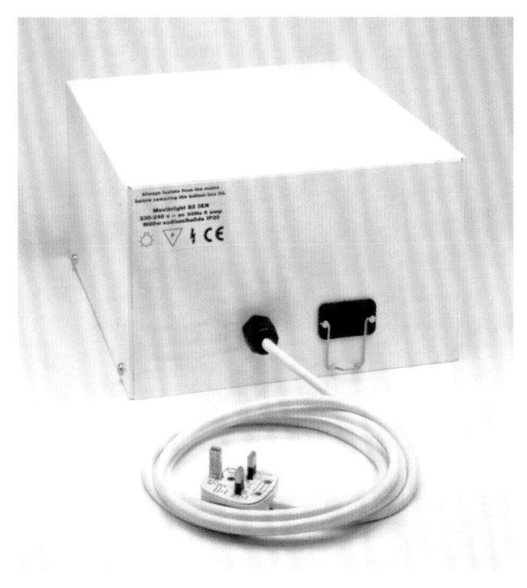

Example of a HID Ballast

Over time, like HID lights, the lamps will diminish in lumen output, so again it is advisable to change the lamps every 9 months. The ignitors also regularly need replacing. In most cases, the ignitor is mounted in the light in such a way that it is very easy to remove and replace.

The lamps are always coated with phosphorescent chemicals which give the lamps a white coloured appearance. The mix of gases contained within the lamp determine the colours emitted. The lamps contain a blend of inert gases, argon, neon, or krypton and mercury vapour, which are sealed within the tube at a low pressure.

Like the high pressure sodium, electricity arcs between two electrodes located at either end of the lamp, which in turn stimulates the phosphor to emit light energy. The light output is greatest in the middle of the lamps compared to the ends. The high frequency lamps tend to take 3-5 minutes to warm up to get to the maximum output, whereas the low frequency takes 2-3 minutes.

The above represents the most popular forms of lighting used in an indoor garden.

Air and Water-Cooled Lights

There are of course more variants available, and are used in conjunction with high pressure sodiums or metal halides. These are the precursors of new innovations in the world of indoor gardening, invented by growers for growers. These are air-cooled and water-cooled lights.

Most growers have difficulties containing the heat of their grow rooms. This is quite honestly an uphill struggle as more growers are bringing in air from outside and venting out hot air from inside. So as you can imagine, during the colder months no problem in most cases, however, during the warmer and hotter months, it can be an absolute nightmare.

The result was that most growers dismantled during the hotter months, reducing the growing seasons considerably. So, through problems arising and then problem solving, air-cooled and water-cooled lights hit the market. And they are quickly becoming the most popular lighting available to date.

Air-Cooled

Air-cooled, like the old Volkswagens, are a clever bit of kit. There are currently two types available – open and shut.

The open air-cooled hood or reflector that hoods the lamps, is designed in such a fashion that an extraction fan or ducting to an extraction fan can be hooked up to it. The hood has been designed to allow the heat emitted from the lamp to escape up through the reflector. As the heat rises, the extraction fan sucks it away and directly from the lamp. So the hood remains cool to the touch and the heat emitted from the lamp is kept to a minimum. These types of hoods also allow for great air flow

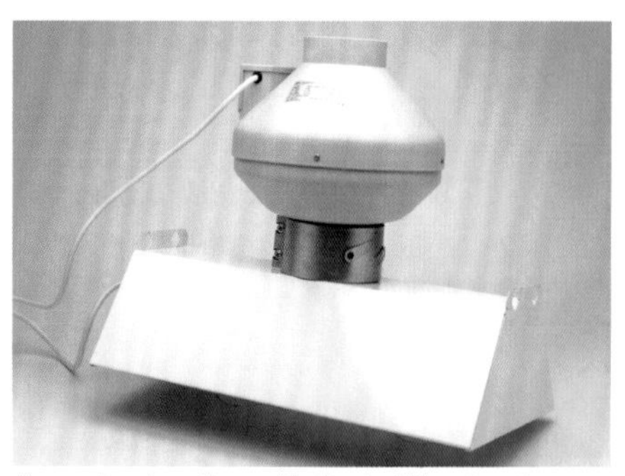

Example of an Open Air-Cooled Hood and Extractor Fan

over the top of your plants due to the open nature of their design. This air flow is believed to be a great contributing factor to the pukka results open air-cooled hoods give. The reduction of the heat from the lamp thus improves air flow, allowing the hood to be lowered more so than a traditional reflector can be. So, in effect, you can get the lamp a lot lower to the canopy of your plants compared to traditional reflectors. This reduces the foot candle depreciation and dramatically improves the lumens available to the plant. Plus, the improved air flow maximises the available CO_2 to the tops of your high energy plants. The open hood also maximises the concentration of light available from the lamp by having the lamp completely enclosed in reflective material, like a shoebox without the lid. This gives you a square of concentrated light. The light emitted does not have any hot, or for that matter, weak spots. Just a solid beam of intense light. In fact, recent tests have proved that this type of hood in general, generates approximately 20% more lumen output compared to other open ended hoods that are available. This does not include the fact that you can get the hood closer to the canopy of the plants without burning the tops.

The closed air-cooled hood or reflector is an entirely different beast. This type of air-cooled hood is available in two different types of design.

The first is a glass tube with a minimal reflector built into it. The reflector is minimal due to the fact that it is mounted within a 6 inch diameter tube so the tube obviously restricts the design of the reflector. The glass tube is completely round with a 6 inch diameter hole at either end of the pipe. The idea is that the hood is isolated from the grow room. So, in fact you have an extraction fan mounted directly to the reflector pulling air from the hood, which can be connected to another hood in turn,

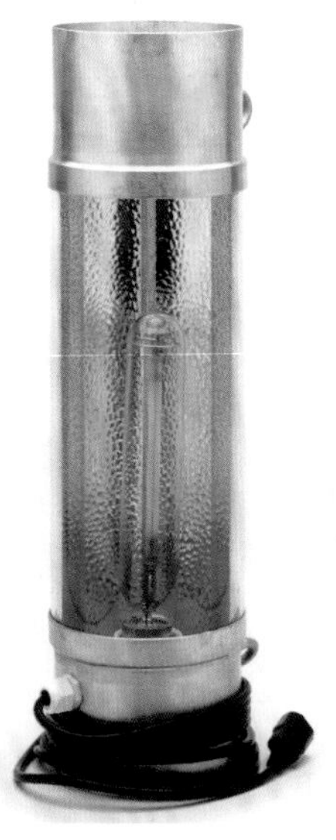

Example of a Glass Tube Air-Cooled Hood

thus linking the lights together to create a network of hoods that are isolated from the grow room. In effect, you have ducting connected to the outside, feeding to the air-cooled hoods, which then connects to the extraction fan which is then vented outside and completely isolated from the grow room. This is a great system to employ if you are injecting CO_2 into the grow room as the lights are generating the bare minimum of heat, and the extraction fan is only venting the lights and not the grow room. Consequently, this reduces the overall loss of CO_2 in your room. The only problem with this type of configuration is that the hoods are firstly enclosed in glass. Curved tubular glass has a tendency to reflect light back in on itself. The result is restricted light output through the curved glass hood. The other contributing factor is that the restricted reflector that is placed within the hood also does not make the most of the light that is available from the hood. Add these factors together and you have a very big disadvantage compared to the open shoebox hood. However, this type of lighting arrangement does allow you to lower the hood very close indeed to the canopy, as the heat generated by the lamps is at a minimum and as a result, the foot candle depreciation pyramid is greatly improved. The plants get good lumen output, but only over a restricted area. As a consequence, you have to use more lights to do the same job. If you have the budget and employ these well, the results are outstanding.

If you are not concerned about injecting CO_2, then growers tend not to isolate these from the grow room. Instead of connecting ducting from outside, they leave one of the hoods disconnected, so when the extraction fan pulls air over the lamp, it also extracts air from the grow room,

The other type of closed air-cooled hood is that of a shoebox design but enclosed with a sheet of glass. The hood is slightly different to that of the open shoebox design as it has an air inlet and an air outlet. So it works in principle just like the air-cooled tubes, but has the benefits of the open air-cooled hood, i.e. the open air-cooled hood only has a hole to remove the hot air; the closed hood has two holes, air in and air out. You can, therefore, link many links together using spigot boxes and manifolds as you can with the standard open-ended air-cooled hood. The downside of this type of hood is the fact that the lamp has another sheet of glass to penetrate through. This, due to the nature of

Example of a Closed Glass Air-Cooled Hood

glass, reduces some of the lumen output of the lamp, however, not to the degree that the curved glass tube does. Overall, there is a slight loss of lumens, but a greatly improved foot candle depreciation pyramid; these lights are almost the best of both worlds. The cost, however, is not cheap. The glass used needs to be heat resistant glass and this type of glass is very expensive indeed and normally costs, in some cases, as much as the hood.

To sum up from the different air-cooled hoods available, the most commonly used and the most affordable is the open-ended shoebox type of hood. However, depending on the different circumstances, they all have their pros and cons, but in any case are well worth the investment allowing you, the grower, more control of your environment all year round.

Water-Cooled

Water-cooled lights are an entirely different beast. They are possibly the most state-of-the-art lighting equipment to come about in the last decade. These unbelievable lights simply do not generate heat. They are basically a normal high pressure sodium or metal halide light with a twist. They use an outer glass jacket to enclose the lamp completely, so you have a lamp suspended within an inner tube of glass. They rely on heat exchange via a large water reservoir. The principle is basic but highly functional. A pump submerged within a large water reservoir, pumps water at a slow rate through the outer glass jacket. The outer glass jacket hoods the lamp with watertight seals. The water, due to its direct contact with the lamp, exchanges heat radiated from the light and the reservoir of water absorbs the heat. The result is a water-filled jacket which you can hold, even after hours of use. They simply do not heat up. This allows the indoor gardeners to get the hood as close as they like to the canopy of the plants. As there is no radiated heat, they simply will not burn the plants. The only thing that could burn the plants is that the lumen output is so great, due to no loss of foot candle depreciation; the literal light output might be too much for your high energy plants to tolerate.

The invention of these lights has heralded a new evolution in super charging indoor gardening. Due to the nature of their design, the generation of no heat allows the environment to be controlled at a level previously unheard of. For example, day and night time temperatures can be easily maintained within a 2-5° differentiation, meaning that the night time temperature is only slightly less than the daytime temperature, thus creating an atmosphere that does not generate possible shock to the plants.

No heat output via the lamps also creates an environment that needs little extraction and allows

CO_2 injection to be maximised and minimal loss occurs due to less extraction. The minimised heat output also allows the lamp to get closer to the plants than has ever been possible before. You can, as stated before, drop these lights on your plants. And remember, double the lux, double the yield. Due to no foot candle depreciation, the yields are quite simply stunning. The downsides are, and there are always some, are as the light is suspended a lot lower to the canopy of the plants, the area covered is less than normal high intensity discharge lighting. Also, as the light itself has to firstly penetrate through a layer of water, then penetrate through a layer of glass, and as the glass is tubular in design, which due to is own nature, reflects light back onto itself, lumen output is less than a standard HID light. However, due to the fact you can literally touch the canopy of the plants with these lights, the lumen output is far superior to normal HID lights, however, to cover the same area you would need more of them.

Water-cooled lights also require more maintenance than regular HID lights. Algae build up is common but easily resolved by adding some H_2O_2, or a product called Pump Clear; either products destroy and stop build up of algae and calcification. The pump also needs regular attention and cleaning, as the pump is perpetually on when the light is on. They also have limitations as most pumps do not have a head or lift of over 1.5 m and this means that if you position your light above 1.5 metres, the pump will not work at its designed capacity, resulting in possible problems with the water-cooled light. Of course, you can get pumps with larger head capacity, however, a larger head normally means a faster flow rate which results in less heat being absorbed via the water exchange, which again, can cause problems.

Example of a Water-Cooled Light

Some people do not use a large enough reservoir to allow sufficient heat exchange. This again causes problems, as the water-cooled lights depend on a large volume of water to exchange the heat, and each light requires a large and, if possible, an independent reservoir which can take up space. However, they can be employed outside the growing environment. Apart from these minor problems and in the right hands, these lights are outrageous. They have built-in safety features, namely a thermo cut-out, so if the heat of the light or reservoir becomes too much, then the thermo cut-out stops power to the lights until the reservoir or the light is cool enough to allow re-ignition.

These lights have been designed by growers for growers and get over the problems of hotter summer months and other possible problems with heat output via lights. You can use as many of these lights as possible in areas where previously you could only use a minimum of HID lights. They do represent higher maintenance, but if indoor horticulture is your thing then this will not cause any concerns. Overall, a great and needed invention.

Photosynthesis and Light

In absolute basic terms, light is food. Everything else we give the plant simply aids to digest and utilise the light. More light equals more yield. The more light a plant receives, the greater it will become providing that all its other needs are catered for. Simply, if the plant does not get enough light, nothing else you use will make the plant grow any larger or produce any more yield.

The main turbo-charging factor in an indoor garden is plenty of available light. Light density of an indoor garden is the determining factor for outrageous yielding, high energy plants.

Lumens, Lux and Foot Candles

The knowledge of lumens or light density is paramount to the indoor horticulturist.

Lumens are a unit measuring light intensity striking an area of surface. Lumens are a measure equal to a number of candles lighting an average book positioned 1 foot away from the light. The simple definition of 1 lumen is equal to the light generated from 1 candle shining on 1 square foot of white paper held 1 foot away from the flame.

So 1 lumen = 1 foot candle or 1 candle of light intensity, per square foot, held 1 foot away.

Now, you have possibly also heard of lux as an expression of light intensity. The metric unit is called the lux and this is simply the amount of light falling on one square metre.

1 lux = 1 metre candle or 1 candle of light intensity, per square metre, held 1 foot away.

A single square metre works out to be approximately 10 square feet, so the lux number is 10 times the lumen number. So 1 lumen x 10 lux is simply the metric way of doing it.

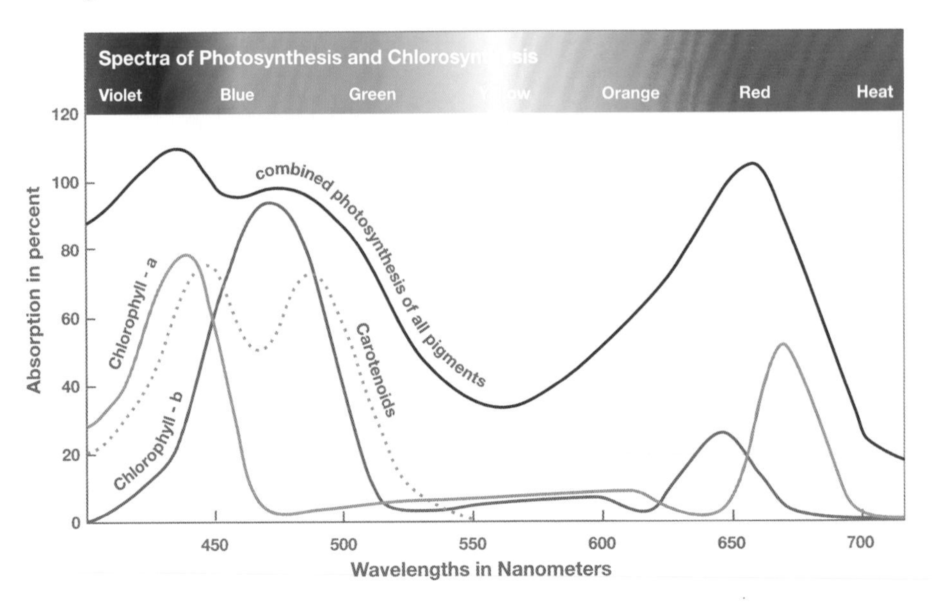

1 lux is only $^{1}/_{10}$th of the lumen because the same amount of light is needed to fall on 1 square foot now has to fall on 10 square feet. So the measurement is divided by 10.

These measurements are a vital tool in growing your favourite plants. The lux output of your lights is needed as the relative wattage, so any light source is very easily confused. The wattage of a lamp can mislead you into thinking that this will be sufficient as all lights are rated differently in terms of lux output.

For example, an incandescent 100 watt standard table lamp shining on a book held one foot away generates 175 lumens onto the page of the book.

If you were to swap this lamp to a 100 watt mercury lamp, the lumen output generated onto the surface of the book will be approximately 600. Replacing this light source with a 100 watt high pressure sodium will generate approximately 1400 lumens on the book. So as illustrated above, it is easy to see that wattage has no relevance on light intensity output.

Different Lights

Different lights generate different amounts of light making certain lights more efficient to use than others. An example is a standard incandescent actually only uses approximately 10% of the energy needed to run it on generating light.

- A mercury blend actually uses approximately 30% of the energy needed to run it on generating light.
- A fluorescent actually uses approximately 40% of the energy needed to run it on generating light.
- A metal halide actually uses approximately 55% of the energy needed to run it on

Getting the Most out of Artificial Lights

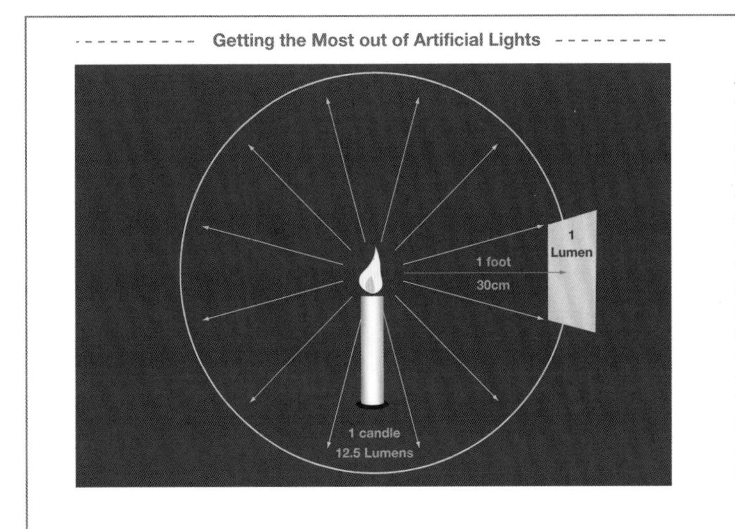

As shown in this diagram, the light projected from a single source is not fully harnessed. **One candle in reality, puts out a total of 12.5 lumens in all directions.** However, only one out of 12 lumens falls onto a square foot of paper held one foot away from the candle. Therefore, only 1 lumen is harnessed whilst the other 11.5 lumens are wasted.

generating light.

- High pressure sodium uses approximately 65% of energy needed to run it on generating light.

The rest of the energy that is not used in producing photons of light is wasted as heat.

At present there is yet to be invented an extremely efficient lamp. And as you can see from above, some are better than others. Although you are still paying for 100% of the energy used, at best this energy is converted into 65% of light and 35% heat.

It is worth noting that for every panel of glass the light has to pass through, an approximate loss of 10% of light occurs. And with this in mind, light passing through a water jacket can lose up to 20% light output, and this light output loss is found mainly in the red spectrum.

It is also worth noting that a 1000 watt high pressure sodium placed one foot away from your book will deliver approximately 140,000 lumens of light. However 10% of lumens is heading for and illuminating your book, the other 90% of light is shining in every other direction in a 360° sphere, so your book is only benefiting from 10% of the lamp's available light. So in effect, the book actually only receives 12,000 lumens. Through the use of good reflectors or hoods, and this is the reason why it is best to buy the more expensive reflectors, is that the 90% of wasted light can be captured and redirected where it is most needed; on top of your beautiful plants.

Now if we go back to the 100 watt incandescent table lamp shining out 175 lumens on your book positioned one foot away. This same light, but being measured from the very edge of the room, will give you approximately less than 1 lumen of light per square foot. The reason for this is distance. Distance dilutes light and dilutes it fast. In short, distance dilutes light density.

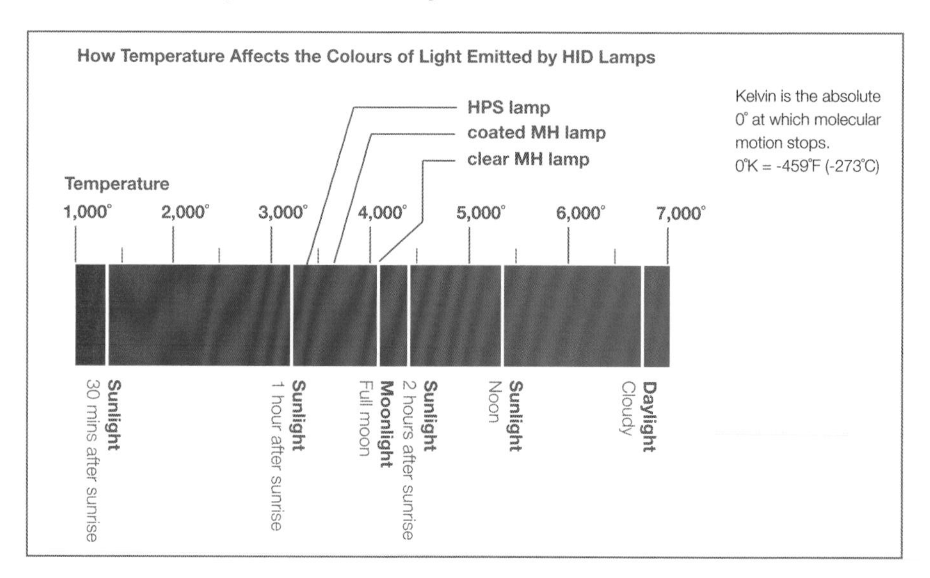

Light Intensity

This page shows an illustration of a square of paper held 1 foot away from the light source, illuminating an area of 16 foot squared. If the light source is the 100 watt high pressure sodium as mentioned above, then you get 12,000 lumens shining on the paper surface one foot away. Spread these lumens over the 16 foot squared area on the floor and you now only receive approximately 750 lumens on each of the 16 squares. This simply shows that light four feet away from the paper's surface is actually 1/16th as intense as it was compared to one foot away. The sad fact is that with every extra foot or 30 cm that light has to travel, the light intensity is reduced by half. Now this is really highlighted in a room that is 8 feet high. If the light is positioned at the highest point, eight foot from the floor, the actual light hitting the floor is approximately 1/64th of its original 1 foot density, making the lumens hitting the floor less than 187.5; a similar light output of an incandescent 1 foot away!

Light Intensity (in Lumens) decreases over distance

Foot Candle Depreciation Pyramid

So, as explained, you can now see the absolute benefit of keeping those lights as close as you can to the canopy of your plants (without burning them please!).

- 1 lumen is 1 foot (0.3 m) from source. Light intensity = 1 candle per square foot (0.09 m^2)
- $^1/_4$th lumen is 2 feet (0.6 m) from source. Light intensity = $^1/_4$th candle per square foot
- $^1/_9$th lumen is 3 feet (1 m) from source. Light intensity = $^1/_9$th candle per square foot
- $^1/_{16}$th lumen is 4 feet (1.3m) from source. Light intensity = $^1/_{16}$th candle per square foot.

Light Requirements for Plants

	Foot candles	Lux	Hours of light
Seedling	375	4,000	16 - 24
Clone	375	4,000	18 - 24
Vegetative	2,500	27,000	18
Flowering	10,000	107,500	12

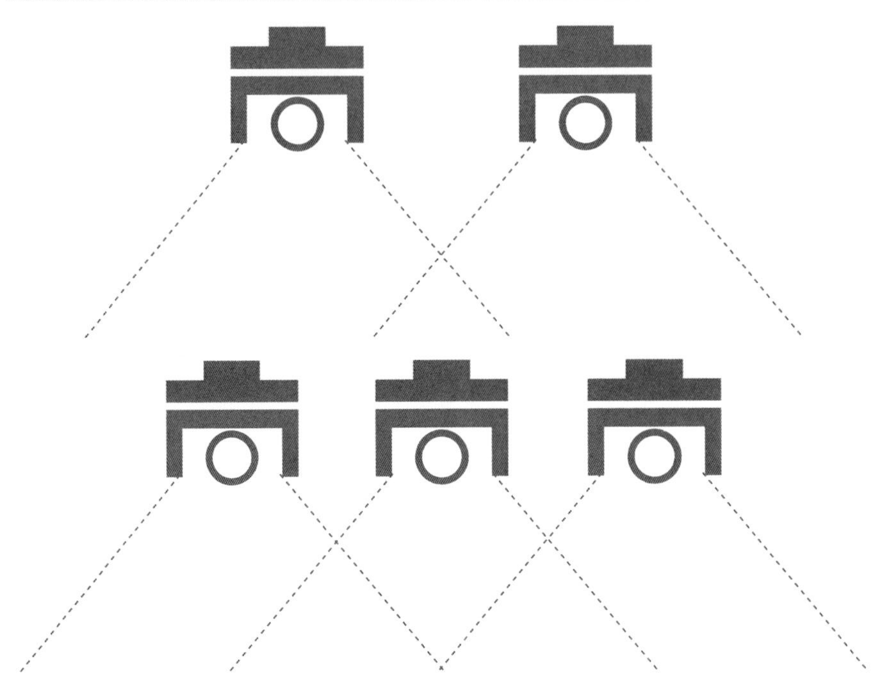

As shown, when using three 600 Watt HIDs in conjunction, they will generate more light and cover a greater area compared to just two 1000 Watt HIDs. Lesser wattage HIDs can also be located closer to the plants' canopy allowing for maximum use of lumen output and therefore, no loss via foot candle depreciation over distance.

Lumens

The goal is to give plants 10,000 lumens

1000 watt High Pressure Sodium
(approximate lumens)

1 foot away	140,000 lumens
2 feet away	35,000 lumens
3 feet away	15,555 lumens
4 feet away	9,999 lumens

1000 watt High Pressure sodium @ 4 feet = 10,000 lumens –
4 x 4 = 16 square feet, 1000 watts / 16 square feet = 62.5 watts per sq ft

600 watt High Pressure Sodium
(approximate lumens)

1 foot away	90,000 lumens
2 feet away	22,500 lumens
3 feet away	9,999 lumens
4 feet away	6,428 lumens

600 watt High Pressure Sodium @ 3 feet = 10,000 lumens –
3 x 3 = 9 square feet, 600 watts / 9 square feet = 66 watts per sq ft

400 watt High Pressure Sodium
(approximate lumens)

1 foot away	50,000 lumens
2 feet away	12,500 lumens
3 feet away	5,555 lumens
4 feet away	3,571 lumens

400 watt High Pressure Sodium @ 2.25 feet = 10,000 lumens –
2.25 x 2.25 = 5 square feet, 400 watts / 5 square feet = 80 watts per sq ft

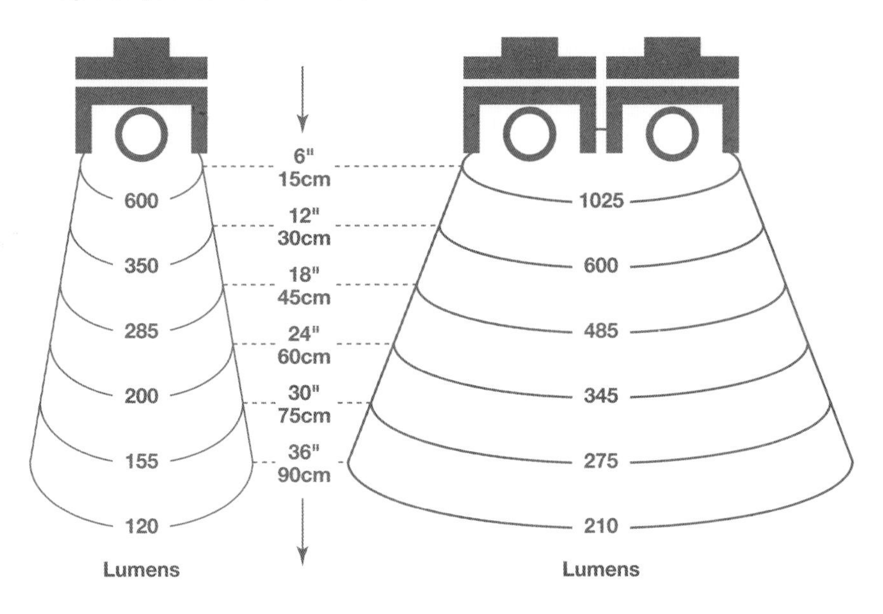

Contactor Relays and Timers

Some growers still do not know how much of a necessity these contraptions are. To explain the need for them when controlling the switching on and off of your lights is simple.

If you are employing a light or lights that are rated at 400 watts or more individually or collectively, then quite simply, if you plug these into a normal domestic 3k timer that you can buy at any hardware store, you will find that in no time at all, your timers do not switch the lights on or off any more. The timers still tick but the switching ability is fused either perpetually on, or perpetually off. As you can imagine, if you only visit your plants when the room is on a daylight cycle and you never go in at night, if your timer blows, the lights can effectively get stuck on and you will never realise your fundamental mistake until you question why your plants are not flowering. Moreover, if this situation occurs in the middle of a flowering cycle, then you will force the plants back into a vegetative state of mind and the end result will be a possible failed crop. Or at best, a crop with which you will have to start the whole flowering process over again.

On the flip side, you might be a gardener that only checks the grow room during the night cycle. So, as you can imagine, if the lights get perpetually stuck on off, then gradually over the course of a week or two, the plants will suffer gradual shock, increasingly getting sicker and sicker as no light comes on, so the plant cannot photosynthesise. If the grower does not make the connection that maybe the lights are not working, the plants will kick the bucket.

Therefore, as illustrated above, it is fundamental that the lights are switched perpetually, and at the same times, on a daily basis.

Domestic Timer Problems

The reason that a 400 watt or greater wattage light will blow domestic timers is not obvious to the layman. The amount of starting current a 400 watt or greater light pulls to ignite the lamp, far exceeds what a domestic timer can handle. This load is only drawn for a few moments until the lamp has been struck and ignites. However, this load repeatedly drawn over the timer for a few seconds on a daily basis, will ultimately blow this timer where the contactors within the timer are fused open, leaving the timer perpetually on, or fused shut, leaving the timer perpetually off. The end result is lights stuck on, or lights stuck off.

The strange thing is that to begin with, the timer functions properly, but over time and this over time bit is variable depending on the wattage of light employed, over time they stop working. In some cases this can be weeks after employing your new timers and in other cases it can be days. One thing for sure is the risk involved in timing your lights properly, outweighs the cost of buying a decent contactor relay timer which can handle the load it requires to fire your lights.

Contactor Relays

There are many contactor relays available to run HID lights. Some HID lights even have the contactor relay and timer built into the ballast itself, however, these types of lights are typically very expensive and some makes of them are prone to go wrong. With this in mind, if your relay or timer goes and it is built into your light, then you have to take your whole light back for repair. Here, if you were to buy the timer and relay separately to the lighting unit, then if the timer or the relay plays up, you still have your functioning light, and the relay and timer can be easily fixed or exchanged with no down time associated with your grow room lights. So it makes better sense to buy the relay and timers separately to your lights.

The smallest contactor relay is a 2k unit. These can power up to two 600 watt HID lights, or a single 1000 watt light or two 400 watt lights or a combination of a 400 watt and a 600 watt lighting set-up. They have two leads that come out of the box with the contactor relay in it. These two leads have domestic plugs on them that plug directly into a twin wall socket. One of the domestic plugs should be labelled timer; this plug should be plugged into the wall via a normal domestic timer. So to recap, both plugs are plugged directly into a twin wall socket, one via a domestic timer. The lights can then in turn be plugged into the sockets that are provided on the contactor relay box. The reason for the relay box to be plugged via two leads into the twin wall sockets, is that both leads can draw electricity via the twin socket simultaneously, therefore, relieving the load pulled via the domestic timer. A contactor relay then absorbs most of the load and the lights fire with minimal stress caused to the domestic timer.

Variations of these can be 3k relays with four plug output. These can run up to three 600 watt HID lights or four 400 watt HID lights, or the equivalent. 4 k relays with four plug output can run up to four 600 watt HID lights or three 1000 watt HID lights, or the equivalent. 6 k relays with 6 plug output can run six 600 watt HID lights or four 1000 watt HID lights, or the equivalent.

All of the above employ a double plug strategy and should be plugged directly into a twin wall socket.

It is advised that you get more sockets than you actually require as you can also use these timers to plug in your CO_2 unit or other equipment that you would employ during the daylight period only. And as most other equipment does not draw much load, it makes this a very tidy way of powering your equipment.

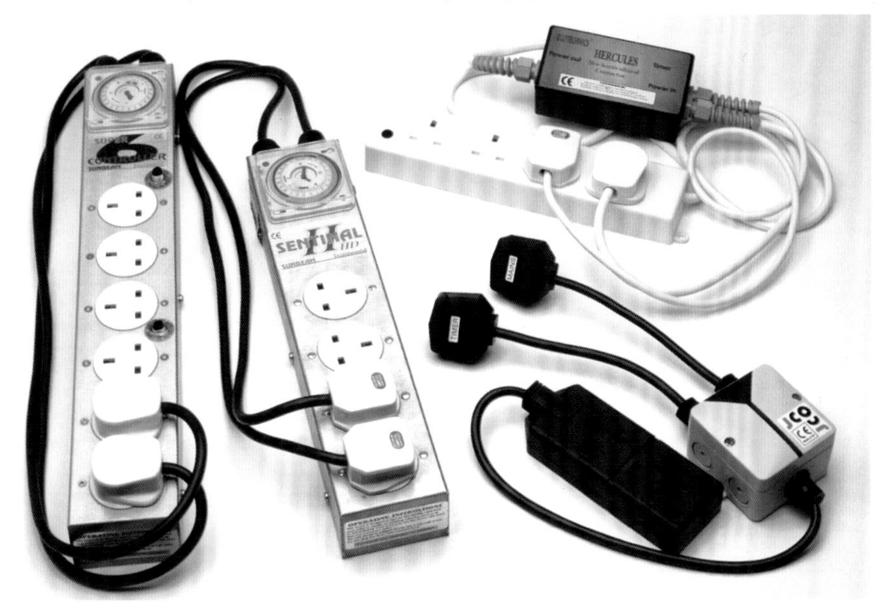

Examples of Contactor Relays

It is recommended that you employ these units if you are using a 400 watt light or greater, however, a 400 watt light is sometimes regarded as borderline in that some timers can just about handle the switching load of a 400 watt light. Nevertheless, this is, as stated before, borderline and most people barring a few exceptions find that a 400 watt light will blow the domestic timer that they have employed. And again, it may not actually be apparent till weeks and in some cases months later that the timer has in fact stopped working. Consequently, you are lulled into a false sense of security believing that you have saved yourself £50 when 6 weeks later in the middle of the 12 hour cycle, your lights get stuck on 24 hours a day. And as you only check the room during the daylight cycle, you are none the wiser until the point that your plants revert back to the vegetative cycle and your prized blooms are lost to an outbreak of foliage. So, if you are broke and you are only using a 400 watt light, then keep an eye on it and be prepared that in most cases you will blow the contactors within the timer.

To sum it up, it is recommended that if you are using a 400 watt HID light or more, then you need this piece of kit. It will buy you reliable switching and timing of your lights and not to mention total peace of mind. Don't be a tight arse – get this bit of kit, otherwise you will live to regret it.

Ozone Generators

New to the market place but already very popular amongst indoor gardeners are the ozone generators.

These machines are primarily sold for their excellent odour controlling properties. They simply destroy odour and odour containing particles via oxidisation, effectively removing any odour from the grow room and anywhere else you put one of these machines.

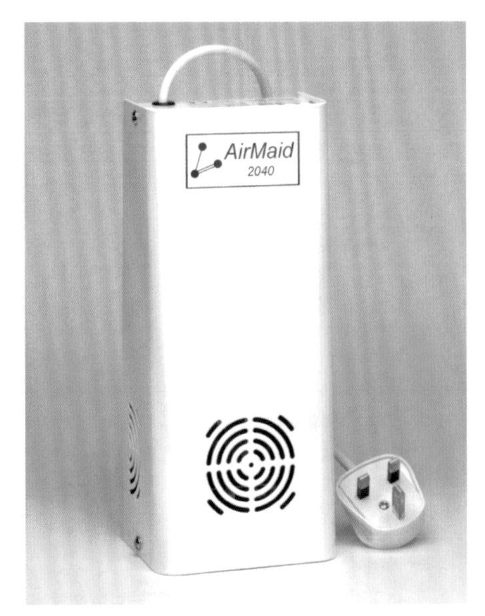

Example of an Ozone Generator

In the right hands, these little wonders are superb, however, with bigger and stronger versions now available, these machines in the wrong hands can also be very detrimental to the plants and even to one's health.

Ozone

An ozone generator makes O_3 (ozone). This is a semi–stable oxygen molecule and this O_3 molecule is attracted to particles to oxidise against so that it can release its third oxygen atom. In effect, the ozone generated via these machines looks for small organic particles to oxidise first, effectively removing any particles relating to smell. However, if the room is not properly ventilated and it has oxidised all the available airborne particles, it will then oxidise anything it can in order to release its third oxygen atom.

This means that if you were to put one of these

machines into a sealed grow room box with no ventilation, over a long period of time it will oxidise against the living plants. Then, given enough time, will turn on the plastics in the grow room. The end result will be no more living plants or hydro equipment to grow them in. It must be said that this can only be achieved if the grow room or box is completely sealed. Adequate ventilation will remove this possibility from materialising.

Ozone generators are also a very effective means of bacteria and fungi control. Ozone feeds on and destroys bacteria and fungi with haste. Outdoor plants have, and always have had help with fungi and bacteria due to the fact that the sun produces UV and ozone which, as stated above, helps control and destroy fungi and bacteria.

Ozone, which is being perpetually manufactured by the sun and is found at the outer atmosphere of our planet, is generated via the sun's far ultraviolet (UV) rays. These UV rays have extremely high energy photons; in fact twice as high as the photons found in orange or red light rays. When these highly charged photons hit the oxygen gas (O_2) in our atmosphere, they then expand the electrons and a third oxygen fills the gap. This then creates O_3, which is simply three oxygen atoms stuck together.

Ozone does not like having these three oxygen atoms stuck together and wants to get rid of the third atom so it can return to its stable state of O_2. The ozone needs to rid itself of the third oxygen as soon as its new energy begins to deplete, which is normally after about 30 minutes. After this time, the third oxygen atom loses energy and falls off to look for a partner. Inside an indoor garden, fungi and bacteria and the nature of some plants, often generate offensive odours. These odours are simply very small particles that float around in the air until you breathe them in. They are then captured via receptors in your nose and mouth; these receptors absorb them, then relay the smell to your brain. Ozone generators counteract these particles and also counter any airborne pathogens, destroying any odour containing particles or pathogens. The ozone which is floating around in the environment of the grow room, due to its nature, is in constant search for a partner. When it gets near a small odour particle, its magnetic charge grabs the particle and neutralises it via oxidisation against the organic bacteria and fungi that cause the odour. The oxygen (O_2) left, is once again stable.

Too Much Ozone

One word of warning concerning ozone generators. If the ozone levels get too high in the grow room and they have oxidised all the particles in the atmosphere, they will start on the plants and if you are in the room, then they also attack you, especially your lungs. As you breathe it in, the ozone will want to oxidise the cells of your lungs. If you are employing an ozone generator and the plant develops bleached patches and streaks, then your ozone levels are too high. Or if you enter a room and you can smell elevated levels of a type of chlorine smell, then again the ozone level is too high. It is recommended that before entering a room to work on your plants, that you should switch off the machine and wait 30 minutes before re-entering the room, just to be safe.

To some, these machines are superb, and if used well you cannot live without them. A complete must if fungi, bacteria or smells are causing you concern. If smell becomes a problem, then an ozone generator will completely and literally destroy all odours, dead.

UV Filters

Inline UV water filters are again a similar product that works in a very similar way to the ozone generators, however, these are designed for water. Reservoir water is pumped via a contained UV lamp and the water is completely exposed to its rays. These UV rays in turn, due to their high energy, eradicate any bacteria or fungi that is present in the water. The end result is bacteria and fungi free water that has been effectively 'cleaned' by the UV rays.

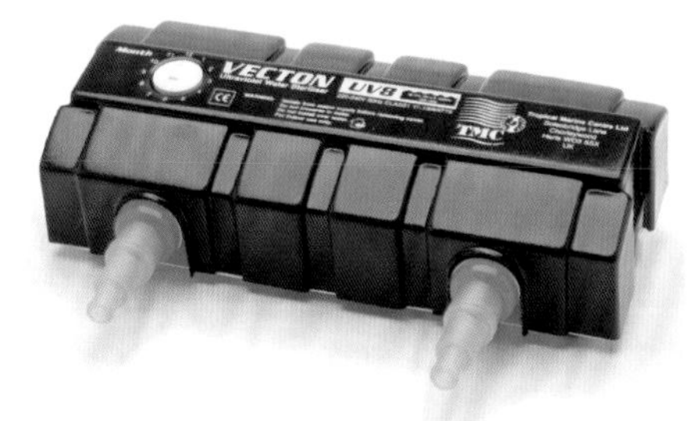

Example of an UV Filter

These machines are simply the only solution to pythium outbreak or root rot; they stop the pythium from breeding, eliminating it from the reservoir which prohibits its spread to other plants. Pythium can simply make grown men cry, and until the release of these products, was almost impossible to control and destroy.

Reflective Sheeting

As stated before, double the lux and you double the yield, so utilising all the available light is paramount inside an indoor grow room. It does not take a rocket scientist to work out that light that hits the walls of a grow room can be reflected back at the plants. Not to utilise this wasted light for the cost of buying the reflective sheeting is simply robbing yourself, but also more importantly, robbing your plants of the extra potential they will gain through more light exposure. The reflective sheeting will also give you another added benefit of lighting areas that the conventional over hanging light cannot reach, which again will help reduce hot spots and give you better all round growth.

Most growers use one of three methods of increasing the reflectivity of the walls of a grow room. These are white paint, black and white plastic reflective sheeting and Mylar reflective sheeting. The ill-informed use tinfoil, however, little do they know that a room painted yellow will reflect more light than a tinfoiled room. Tinfoil is simply one of the worst reflective materials out there.

Matt White Paint

If you are on a budget, then paint the room white. This must be matt white as this reflects a lot more than gloss white and through painting the room matt white you will have made the walls approximately 70-80% reflective. This is cheap, convenient and is quite effective. It is important to keep the walls clean to maintain maximum reflectivity.

Black and White Sheeting

The next inexpensive method is black and white reflective sheeting. Again, just to point out the obvious, the black side goes on the inside so only the white side is visible inside the grow room. You

Examples of Reflective Sheeting

may mock me for pointing this out, however, there are some that have asked which way round it should go, and if you are one of these punters, then I would advise you take up another less taxing hobby. ☺

Anyway, if you employ this method, you will raise the reflectivity of the walls by approximately 80-90% – a good investment for the very little it costs. Again, keep it clean to maintain its reflectivity.

Mylar

The last and most popular, but also the most expensive method, is space age Mylar reflective sheeting. Due to demand, there are now 3 different versions of this product. All expensive, but now even more so, which can only be a good thing for the punter, and the retail industry! The cheapest of the three is bright shiny on one side and dull shiny on the other – again shiny bright side out please! This type of Mylar is not completely light-tight but does reflect approximately 90-95%; the next is a lot thicker and is baked in white, so shiny bright one side and white on the other, I do not need to tell you which way round this goes! – this stuff is 94-97% reflective, it is also completely light-tight unless you puncture it, and is simply the most reflective medium on this planet. The next and most expensive and absolutely brand new is ADF Mylar reflective material. This is slightly less reflective than the other two, however, has a unique heat reflective ability. It completely blocks infrared. This type of Mylar will block any heat signatures given off by your equipment or your grow room. In effect, it highly insulates your grow room and does not allow infrared transmission. It will, however, increase your grow room temperature by up to 30%, but, it will completely stealth insulate it! When employing any of the above Mylars, you can increase the light levels in your grow room by approximately 25%, which is a lot of potential yield that no grower can miss out on.

Mirrors

One more big misconception is that mirrors are great at reflecting light. Although there is some truth in this, the glass itself absorbs much of the usable light. The way a mirror works is that light has to pass through the glass to then hit the reflective material, then the light has to again pass back through the glass to reflect back on to the plants. The fact the light has to pass through the glass twice, restricts the reflectivity of the usable light a plant can absorb. So, in effect, mirrors are not a good choice of reflective medium.

The conclusion is, if you do not make the most of the light that you are paying for, you and your plants will suffer. Reflect it and all will be better.

Light Movers

These were primarily designed and invented to mimic the sun rotating through the sky. The sun is never in the same place, as it is in constant motion, as with a light suspended on a light mover. They were also invented to lower the running cost of your lights by making your light work harder for your money, as a light on a light mover will cover more area than the conventional static light. It was also noted that due to the fact that the light is moving, you could get the light a lot lower to the canopy compared to a static light, without detrimental effects to the plants' health. The lower the source to the canopy, the higher the lux, the bigger the potential yield. One more benefit from using light movers is that it also stops any hot and weak spots in the grow room.

Light movers are a motorised device that literally move the lights back and forth or in circles above your plants. These movers make light distribution more even, and more constant. Static lights tend to generate hot spots and weak spots; some areas of the grow room get abundant growth and other areas get weak and retarded growth – a major cause of this is weak and hot spots. Light movers generally iron out any hot or weak spots that are commonly associated with static light grow rooms. Light generated from a static light is always shining at the same intensity in the same area or spot, and when upper foliage shades lower leaves, the growth of the plant can slow and can cause uneven growth. If the light is in motion, it then bathes the plants in light from all angles, therefore, more light overall is allowed to be absorbed by more foliage on the plant. The end result is better and more even growth. With this in mind, the plants actually follow the light back and forth, and this action of the plants swaying to the light source, as in nature, actually strengthens the plants cells and makes the plants more robust and stronger, allowing for the plants to hold more weight on its strengthened structure.

Currently, there are two different types of mover available – the linear and the circular.

Linear Light Mover

The linear light mover works by tracking back and forth over a distance of approximately 1.5 metres; the good ones move very slowly and at the end of each run, pause for a period of time. This allows for

Example of a Linear Light Mover

the fact that the middle plants, if there was no pause, would receive twice the light the end plants would. So in effect, the light pauses at the end of each run to make sure that all the plants receive the same amount of light. You do need to watch this, as the light remains stationary for longer at the ends, so if the light is too low, it is easier to burn the end plants, but not the centre ones – so do keep an eye on them.

Circular Light Movers

The next and not as popular is the circular light mover. These rotate either in complete never-ending 360° circles, or 180° arcs and are available in one, two, three and four arm versions. These units are great if you are thinking of mixing in some metal halide lights with your high pressure sodiums, which is again more consistent to what the sun would produce. They are, however, very expensive and are high maintenance, so be warned. Nevertheless, they are very effective and are possibly the best thought about light movers on the market. One cheaper version of this circular mover is one that only rotates in a 180° arc. Now,

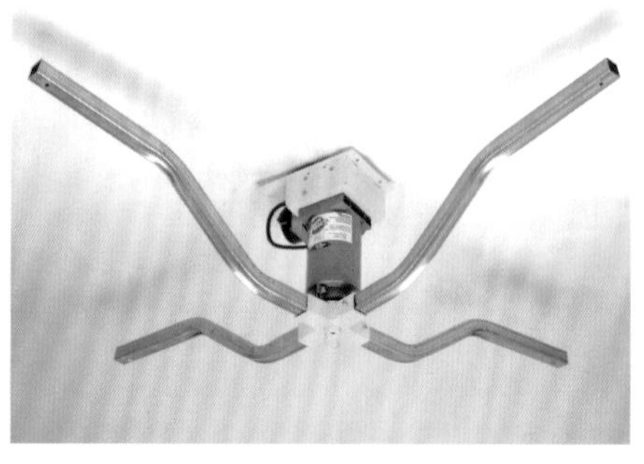

Example of a Four Arm Circular Light Mover

although these are less expensive and are less likely to go wrong, they are in my opinion not as good as the 360° circular light mover or as flexible as the linear light mover. As these only swing in a 180° arc, they are no good for putting different lighting combinations on as only one half of the grow room will get the metal halide and likewise, only one half of the grow room will get the sodium, and is therefore not so practical.

Practicable

If a grow room is set up properly with light movers, they will on average cover up to 25% more plants with greater potential of yield, as more of the plants are bathed in light.

Also, if you are utilising light movers, you are reducing the running costs of your grow room but equally, you are in turn reducing the heat of your room as less lights are covering more plants and generating less heat – again up to 25% less heat.

Employment

As these movers are in motion, it is strongly advised that you use a minimum of 600 watts of light on each arm. If you employ lower wattage lights than this, the plants that are not directly under the light may suffer as the foot candle depreciation from the furthest plant will be greatly diminished. If your light is not powerful enough, then this period of low light will greatly reduce the potential of your plants. So, always use the most powerful light on light movers. That is unless you are raising seedlings or clones as these do not need the same levels of light as an adult plant would.

In conclusion, these tools are well worth the investment as they benefit the plants, the grower and the grow room, however, they are not cheap. Typically a 600 watt high pressure sodium complete light is cheaper to buy than a light mover, so if electricity or heat is not a problem, then spending the money on an extra light will generate you double the lux therefore, double the area or yield compared to only 25% more coverage or yield provided by a mover. If you are a serious turbo charged grower, then you would do both. ☺

Hydroponics
Indoor Horticulture

Chapter 16

Grow Rooms

Most people will set up a grow room in a free spare area like the loft, basement, shed, under stairs cupboard, so on and so forth. These rooms or spaces are often rooms or spaces that pre-dictate the overall size and shape of the grow room. So in general, it is a matter of adapting that by which the space dictates. Whether you are custom building a grow space or adapting one, you should still consider the same points. These main areas of concern are:

- Access
- Electricity
- Ventilation
- Water
- Temperature
- Ease of use

Access

The grow room or space should allow easy access, both on entering and leaving the growing area. Access should not be readily available to most other inhabitants of your house as many dangerous chemicals are kept in most grow rooms, so this room should definitely be out of reach of children, and for that matter, any other unwanted guests. If the door can be locked or even stealthed, all the better. False doors are always fun and quite easy to install and set up. It is also important, if your grow room allows it, to have space for manoeuvrability for you and your plants.

Electricity

It is highly recommended to run a separate ring main straight into the growing space; this is really a job for an electrician and is not too costly. The sparky will also put in a RCD circuit breaker which will give you peace of mind for you and your plants. It is also highly advisable to run more sockets in the room than you actually require as it is very easy to underestimate the amount of sockets one needs for setting up a good grow room; this also will leave you room for expansion in the future. The sockets are recommended to be run around the top of the grow room and not at the normal floor level. This will ensure that the water gets nowhere near the electricity and gives you room for error without the concern of tripping your house or shocking yourself. The help of a professional is worth the extra expense when doing the above.

Ventilation

Ventilation will no doubt be required at some point during your grow, so getting air into the grow room and out again would be worth identifying before you start to deck out the room. Most grow rooms need to extract out the hot stale air; this is done via an extraction fan which sucks air from the room then blows it out via ducting. So being able to extract out to the outside or to the attic, or for that matter even another room, would be advisable. The air simply needs to be able to be pulled out of the grow space and put elsewhere, and the ability for a grow space to have this function is paramount.

Water

Considering the accessibility of water to, and also from your grow room, again is an important consideration to ponder. Although with the availability of easy clip on hosepipes and water pumps, this is easily rectified if you do not have water or drainage in the location that you have decided upon. However, make sure that you have the equipment you require to get water to the grow room, and also to get water from your grow room.

Temperature

Daytime and night time temperature of the grow room should also be considered. Try to avoid areas that get hot during the day and for that matter, too cold during the night. However, again, with the aid of the current technology available to the indoor grower, this can be rectified, but at a cost; that is buying the kit and also running it.

Ease of Use and Environment

Ok, now ponder on this: the area that you are considering using – is it larger than the actual area you require, or smaller than what you really would like? If it is too small for what you really want, do you mind downsizing or should you really be looking at a more forgiving area? If the area is too big then you should think about the possibility of building a grow room inside of the space, or would the extra space be beneficial for working in, or for that matter, future expansion? Both approaches have their pros and cons. So, before you do all that work, give it some serious thought.

If the grow room needs to be built or adapted, then obviously some level of construction will be needed and many trips down to your local builders' merchants. With this in mind, if you are undertaking serious modifications like walls, ceiling, floors then make sure that the materials are up to the job and are capable of handling hot and wet, humid conditions.

Light pollution from outside and inside is another very valid point to weigh up. If you are in a 12 hour cycle, then absolute darkness needs to be maintained to ensure that the plants flourish. So, any possible areas where light can get in or for that matter out, should be sealed and blocked up so light pollution is not possible.

It is also recommended that you ensure that your floor is secure as many hydroponics techniques depend on large volumes of water, and it only takes a reservoir that holds 250 litres of water to equate to a quarter of ton in weight. So this type of weight coupled with getting the floor wet could literally bring down the floor/ceiling. Therefore, ensuring that the floor is secure is of utmost importance and for that matter, it is also recommended that you employ a membrane to keep that floor waterproof.

Noise is also worthy of concern and making sure your room is well insulated helps both with noise and with environmental control. Vibrations should also be highlighted and dealt with. You might have the quietest grow room on the planet, however, one constant, steadily repeated vibration is all it takes to upset you or your neighbour. All vibrating equipment can be isolated by using bungies, rubber backed carpet, or some foaming etc.

If pets and children and other folk do have access to your grow room, then invest in a locked safe for holding all you aggressive chemicals and other non child-friendly bits and bobs.

If you are utilising an already used space, then clear the whole room before use i.e. get rid of the carpets, the curtains, etc. These items will be heavily laden with bacterial and other mould-encouraging pathogens and if kept in a hot, humid room, you will have a cocktail for mould and fungi. Moulds and fungi simply encourage more mould and fungi, which over time will consume your beloved crop.

Grow Room Set-Ups

It is now time to illustrate a few typical examples of different grow room set-ups. Below, you will find an example of the very basic, to examples of the very complex and professional. All grow rooms work, but to differing abilities and convenience.

250/400 Watt Grow Room

The 1-3 plant room. This room is typically favoured by the Percy Throwers, or as a mother plant room and propagation room.

The room is approximately 1-2 m^2 and the walls are lined with reflective sheeting like black and white or mylar, depending on what you can afford.

The room has adequate vent holes at the top and bottom of the grow room or box. Allow at least or approximately a 6 " diameter hole or the equivalent in holes at the top and at the bottom of the grow room or box. Ideally, these should be opposite to each other so the cold air travels across the grow room when the hot air escapes. It is highly recommended that you employ an extraction fan like the lti 100 mm fan to suck the hot rising air out of the grow room and that this fan is mounted at the highest point in the room or box.

This type of grow room typically uses a 250 or 400 watt metal halide if it is a mother and cloning room, alternatively, a 250 or 400 watt high pressure sodium if it is a 1-3 plants grow to finish room.

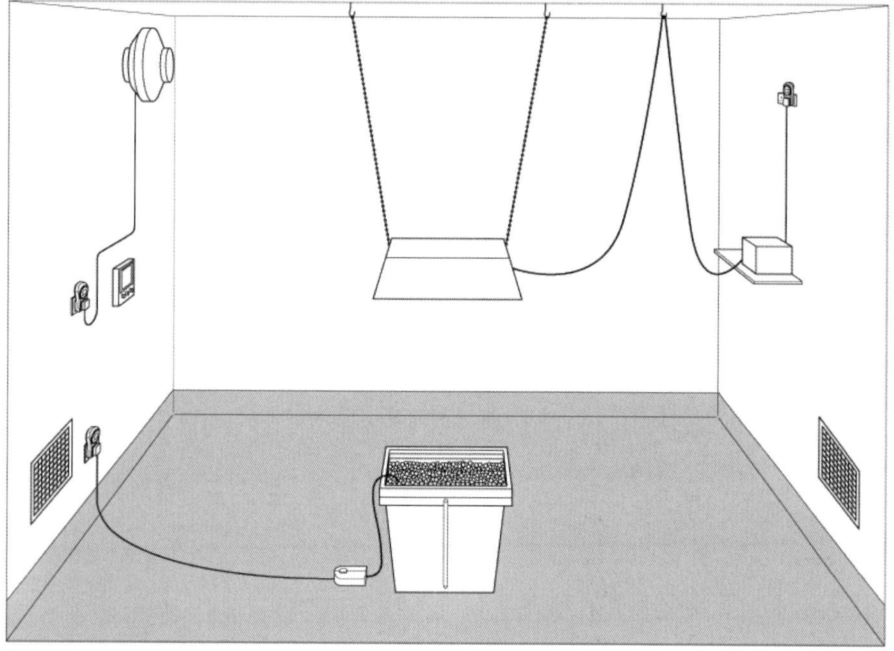

Example of a 250 or 400 Watt Grow Room (1 to 3 Plant Set-Up)

The 250 watt light can be plugged directly into a normal timer and no contactor relay is needed for a light of this power as a 250 watt does not draw enough to stress a typical domestic 3k timer. The light is hung using hook plates and chain from the ceiling of the grow room or box. It is recommended that the remote ballast is placed on a shelf dedicated for that purpose and that it is not left on the ground, as typically the floor of a grow room can get wet, so anything electrical should not be placed at ground level.

The floor should be covered in a plastic membrane to allow for any spillage or leaks that might occur during the set-up or the grow. This membrane should be lapped up at least 4-6 inches up the sides of the grow room or box to contain possible big spills.

A min-max thermometer should be hung on the wall at approximately the same level as the canopy of the plants, as should a min-max hygrometer.

A large drip irrigation water farm would be a good little hydroponics technique to start with and is ideal for mothers or finishing plants.

A propagator can be placed to the side of the hydroponics unit if you wish to propagate cuttings as well.

400/600 Watt Grow Room

The 4-12 plant grow room is used to veg, fruit and finish your plants, and again this is another favourite of the grower.

You would need an area of 2 m^2 or over, and adequate vent holes should be allowed for. It is recommended that you employ a lti 125 mm extraction fan and that this fan is mounted at the very top of the grow room space. Most growers growing this amount of plants fit a 125 carbon filter to the extraction fan to clean away any unwanted odours that the plants may produce. This fan and filter combo is then ducted outside the grow space. Allow at least 150 mm diameter of air intake vent holes; these should be at the bottom of the grow room or spaced preferably opposite the extraction fan, so cool air is forced over the entire grow room.

This type of grow room will require a 400 or 600 watt HID high pressure sodium light. A lot of growers would employ the planta or the agro lamp, so the one lamp will cope quite happily with both stages of grow and bloom. The 400 or 600 watt lighting unit should be plugged into a timer via a contactor relay unit. The ballast should be placed on a shelf dedicated for this purpose and mounted high so that all of the heat generated from the ballast rises straight up and gets extracted out, but also ensures that no water goes anywhere near this piece of kit.

The walls should be lined in mylar reflective sheeting.

The floor should have a plastic membrane put down which should lap up the walls to catch any unwanted spills, big or small.

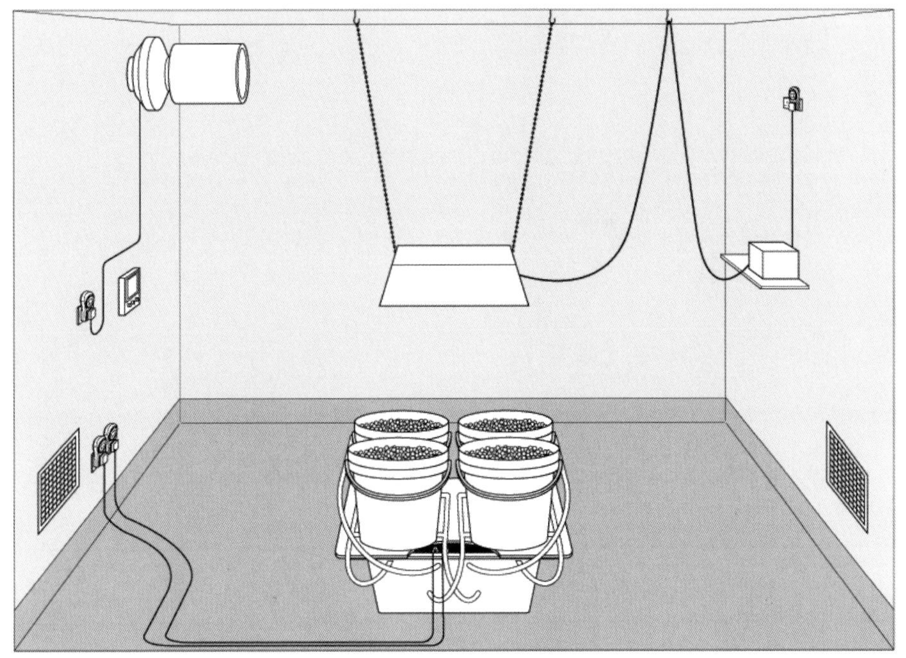

Example of a 400 or 600 Watt Grow Room (4 to 12 Plant Set-Up)

A min-max digital thermometer and hygrometer should be hung at the approximate finishing height of the plants.

A four pod flood and drain system would be a great hydroponics technique to employ in this grow room as it can grow 4-12 plants in this size of space, so the choice becomes yours of how many plants you want to grow.

A propagator can be placed to the side of the hydroponics unit if you wish to propagate as well.

Twin 600 Watt Grow Room

The 16-48 plant room is used to veg, fruit and finish your plants and again is a very popular configuration for a grow room.

An area of approximately 3 m^2 would suffice. A lti 125 mm input fan and a lti 160 mm extraction fan would be a good combination to use to ensure good air exchange. It would also be advised to allow at least another two vent holes at the bottom of the room to maximise air input. The lti 125 mm would be used to push fresh air into the grow room and the air vents to complement this fan. The lti 160 mm fan is best used solely for extraction and should be coupled with a carbon filter to suit the air flow of the fan. It would also be advised to wire both fans to thermostats to ensure the fans are only on when the temperature is too warm. The lti 160 mm output fan should be mounted with the carbon filter at the highest point of the grow room.

It would also be advised in this grow room, to employ an oil-filled radiator also wired to a thermostat, to turn on if the temperature in the night cycle falls too low.

The walls should be completely mylared from top to bottom. ADF heat repellent mylar should be used to cover the ceiling and corners of the walls.

The floor should be covered with a waterproof membrane to ensure no awkward spills. This should be lapped up the wall to catch even the biggest spill.

The two 600 watt high pressure sodium hoods should be hung from the ceiling and the ballasts

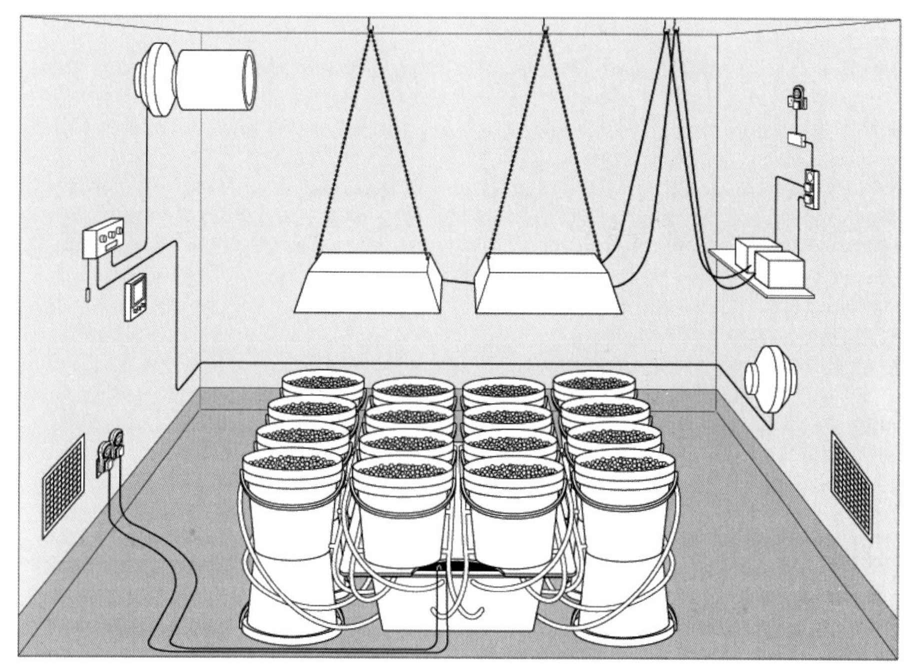

Example of a Twin 600 Watt Grow Room (16 to 48 Plant Set-Up)

placed on shelves mounted high in the grow room. The two lights should be plugged into a twin contactor relay box rated for two 600 watt HID lights.

A min-max digital thermometer and hygrometer should be mounted at the approximate finishing height of the plants.

A 16 pod deluxe system as a hydroponics technique would be an excellent choice to run under 2 x 600 watt lights.

A constant check CF and pH meter would be a good choice of digital meters to use to monitor your reservoir.

A water heater to take the chill out would be a good optional extra.

For greater heat control over the warmer months, as an optional extra, you can upgrade the hoods on the 600 watt lights to the air-cooled variety. This can be of great benefit during the summer when temperature becomes a problem. Basically, the hoods would be connected via ducting to a spigot manifold box which would have a lti 150 mm fan mounted to it. This fan would then be ducted out of the grow room. This additional upgrade is well worth the extra cost when the temperatures rise.

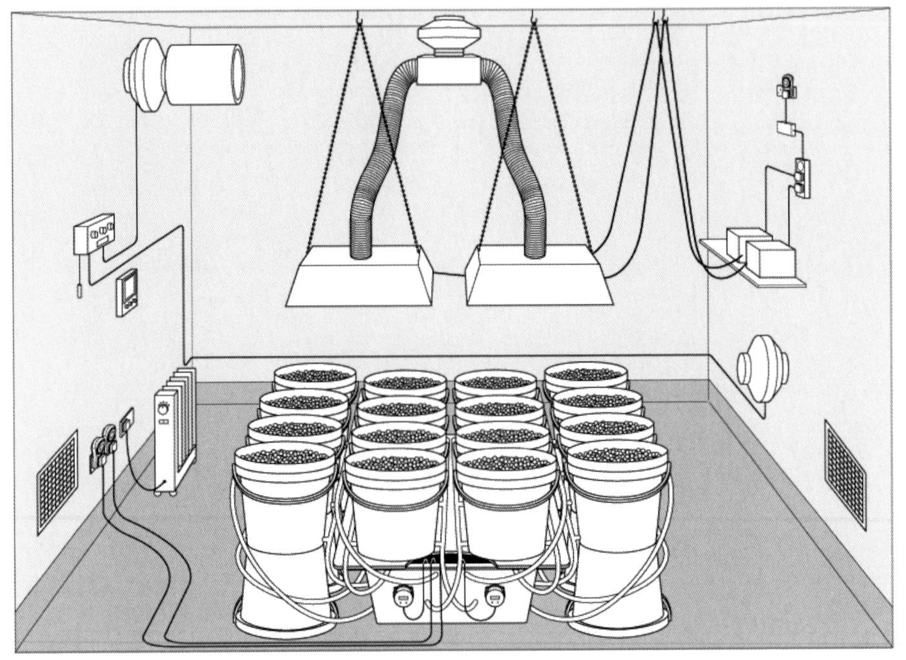

Example of a Twin Air-Cooled 600 Watt Grow Room (16 to 48 Plant Set-Up)

Quad 600 Watt Grow Room

This type of grow room set-up would accommodate 24-72 plants.

An area of approximately 5 m^2 is a sufficient area to locate this set-up in. A 24 super pod system is perfect for this hydroponics technique. The super pod employed uses a massive 220 litre reservoir, which when you are feeding 24-72 plants becomes a blessing, as the greater the volume of water in the reservoir, the greater the stability of the pH and CF is, so in effect the less maintenance you need to do. This system is very modern, modular and versatile plus it also gives you great headspace to grow very big yielding plants. A 24 pod turbo charged aeration system is also advised to maximise the oxygen to the root systems. These large air pumps should be mounted at the very top of the grow room.

A lti 150 mm air intake fan would be recommended coupled with a lti 200 mm extraction fan and carbon filter combo for air out. A couple of extra vent holes low down would also be of benefit to ensure good air exchange. Both fans should be wired to a Prima thermostatic fan controller. This unit will turn both fans on full blast when they are needed, but also when the temperature is controlled, both fans are switched onto an ambient setting to ensure constant air flow, but at a minimised level. So basically, when the grow room is too hot, both fans turn on full blast to quickly maintain the grow room at the right temperature. Once the fans have done the job and the temperature has dropped, then both fans turn onto a dimmed down ambient setting so air flow is constant, but minimised. The fan and filter combo should be placed at the highest point in the

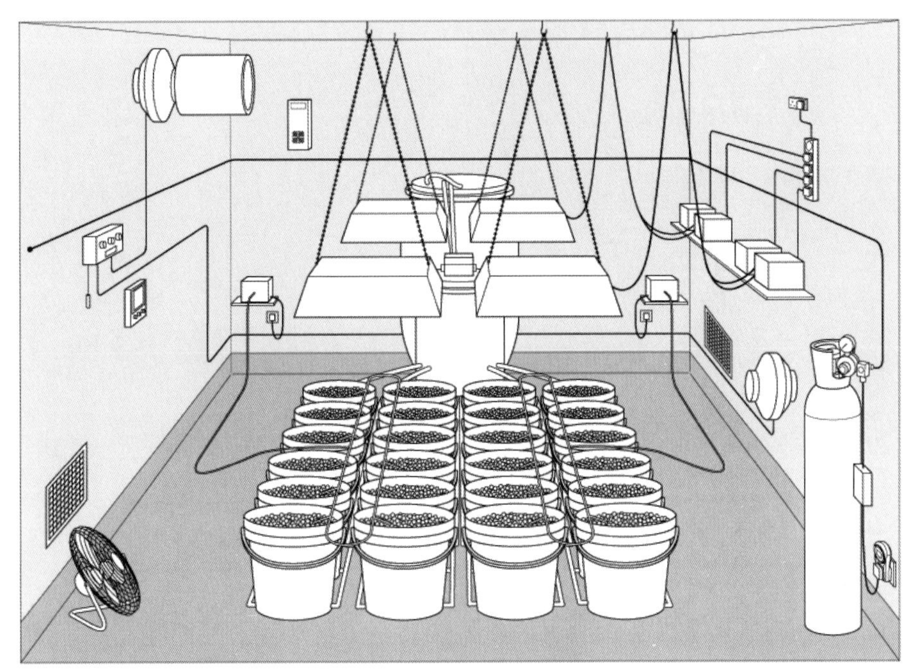

Example of a Quad 600 Watt Grow Room (24 to 72 Plant Set-Up)

grow room.

The four x 600 watt lights should be controlled via a 4 or 6 kilowatt contactor relay timer board to ensure proper and reliable timing of the lights. This relay board, and the ballasts for the lights, should be mounted high in the grow room for heat and safety reasons.

A digital min-max thermometer and hygrometer should be mounted at the approximate finishing height of the plants.

The walls of the room should be completely mylared and the ceiling should be lined with heat signature blocking ADF mylar. The floor should be lined with a plastic membrane liner, which should lap up the walls by approximately a foot.

A constant grow check pH and CF meter would be a good choice of digital reservoir meters. These are constantly kept in the reservoir and measure the CF and pH of the water to always ensure the right levels. These meters can even be programmed to flash if the reservoir falls out of the set parameters. So, if your pH or CF levels get too high or too low, the monitor will let you know so you can make the relevant adjustments.

Oil-filled radiators should be used for the night time wired to a thermostat, so if the temperature falls too low, the heater will turn on to ensure that the room never gets a chill.

Water heaters should also be placed in the large water reservoir to make sure that the reservoir's water is never chilled, but also never warm.

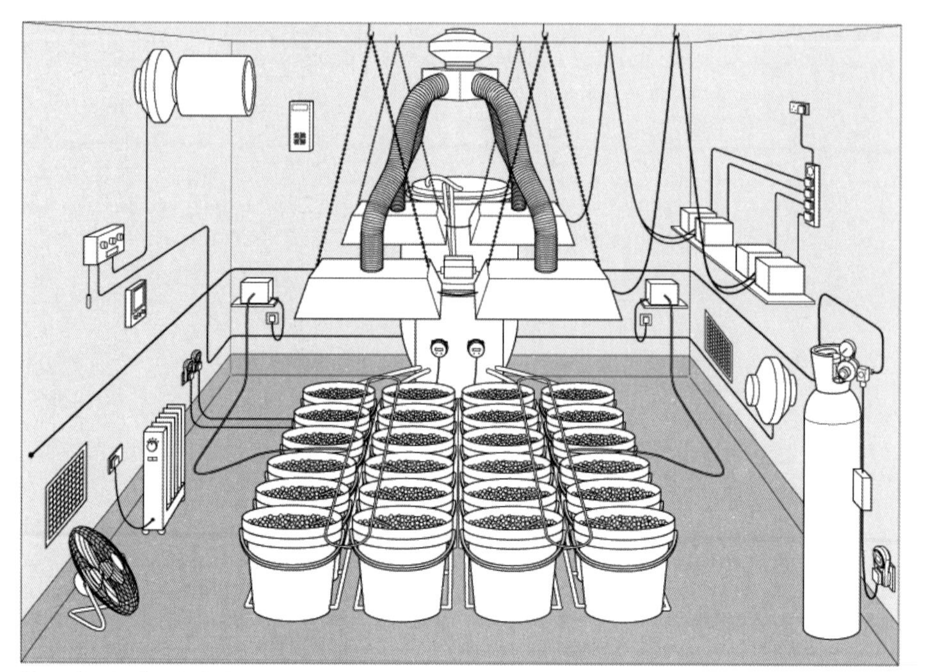

Example of a Quad Air-Cooled 600 Watt Grow Room (24 to 72 Plant Set-Up)

An industrial floor mounted fan should also be used to strengthen the plants and ensure good air flow. An ozone generator would also be recommended to control any extra unwanted odours.

As an additional extra, you should employ a CO_2 bottle and regulator to enhance the levels of CO_2 in your grow room and with this many plants and this much light, it would be highly advisable.

For greater heat control over the warmer months, as an optional extra, you can upgrade the hoods on the 600 watt lights to the air-cooled variety. This can be of great benefit during the summer when temperature becomes a problem. Basically, all four hoods would be connected via ducting to a spigot manifold box which would have a lti 200 mm fan mounted to it. This fan would then be ducted out of the grow room. This additional upgrade is well worth the extra cost when temperatures rise.

Hydroponics
Indoor Horticulture

Chapter 17

The Past Shaping the Present – A History of Hydroponics

The Past

The roots of hydroponics are as old, as deep, and as strong as the understanding of the plants that grow in them. The term "Hydroponics" was first coined by Dr W E Gericke around the 1920s and 30s to define his work with plants. The word is an amalgam of the two Greek words "hudor" meaning water and "ponos" meaning labour which literally translates as "water labouring", or more generally "water works". Whilst the latter half of the 20th century saw the largest leaps in technology, instrumentation and the use of the science, the earliest recorded instance of hydroponics is possibly the almost mythical story of the Hanging Gardens of Babylon.

This wonder of the Ancient world is regarded as one of the first documented, working, hydroponic gardens but the science and development of the idea of hydroponics has been around in one form or another since or before the building of the Egyptian Sphinx and Pyramids. Some researchers have concluded that the Hanging Gardens of Babylon were a very large and elaborate hydroponic system, continually pumping fresh water, rich in oxygen and nutrients, to support its thriving plant life. This is still only supposition however, and the debate continues and no doubt will do so for a long time.

In some of its earliest incarnations, hydroponics followed the general rule of thumb for most scientific advances and breakthroughs which is that the problem forces a solution and the solution becomes an evolution. An example of this is that of the Aztec tribes of Central America. They were forced to feed their increasing population living on the agriculturally void and marshy shores of Lake Texcoco which later became the foundation of their capital city – Tenochtitlan. As a burgeoning nation, they were forced to derive a living from this inhospitable land by their hostile neighbours. Any option of moving to a friendlier climate and neighbourhood was not available to them. Ingenuity and necessity forced them to get creative and invent a whole series of floating gardens. They fashioned small rafts from reeds and rushes lashed together with tough roots; these rafts were called chinampas. They covered these floating planters with soil dredged up from the shallow lake bottom which was rich in a variety of organic debris and decomposing material that over time released large amounts of nutrients. The roots of these plants pushed down towards the source of water, growing through the floor of the rafts and down into the water. They would then plant vegetables, herbs, flowers and even trees on these small floating rafts. As their endeavours grew, so did their gardens and their ambitions. They would then fasten these floating planters together to create huge artificial floating islands of produce; at times these islands reached sizes greater than 200 feet in diameter. Some chinampas even had a purpose-built dwelling for the resident gardener and on market days these gardeners would pole and pitch their chinampas close to the market, harvesting and selling the vegetables and flowers directly to the customers. The ingenuity, creativeness, motivation and hardiness of the Aztec people eventually allowed them to create a great and powerful nation which then went forth and pushed back their hostile neighbours, which allowed them to rule in dominance. They excelled in warfare and the sciences, developing an intricate calendar, medicinal practices and mathematical formulae that were far advanced for their times. All, no doubt, derived from the ability to feed themselves from their simple but highly effective floating hydroponics gardens. Even after their rise and all of their expansion and conquest, the Aztecs never gave up their home on the marshy shores of Lake Texcoco, and more importantly their fabulous floating gardens.

When the conquering Spaniards, on their arrival to the New World in search of gold, witnessed the sight of these floating islands, they were simply dumbstruck. The spectacle of entire orchards of trees somehow suspended but mobile on the water must have been mystifying and possibly frightening to the 16th century minds of the Spanish conquistadors.

The historian William Prescott, who chronicled the downfall of the Aztec empire by the conquering Spaniards, describes the chinampas as "wandering islands of verdure, teeming with flowers and vegetables and moving like rafts over the water". These chinampas were still used on the lake well into the 19th century, albeit in greatly reduced numbers.

Ancient heiroglyphs in Egypt also provide proof of gardens existing along the banks of the Nile with plants growing without soil. Theophrastus (372-287 BC), was a Greek philosopher and Aristotle's successor as head of the Peripatetics. Although Theophrastus's influence and genius was varied, one of his most interesting discoveries was that made with plant life. Theophrastus undertook many studies and experimentation into the growth processes of plants and the environmental effects on them.

Rice is a classic example of a hydroponically grown crop. Its harvest dates back from time immemorial.

However, all of these examples, while exhibiting the basic principles of a hydroponic garden, do not prove the use of a refined nutrient supply. Documented research into the constituents of plant make-up and nutrient requirements for plant growth is not evident until the 1600s.

Jan Baptista van Helmont (1577-1644) was an exceptional and brilliant Belgian chemist, physiologist and physician. Although he tended to dabble in esoteric mysticism, he was an exacting observer and experimenter. Helmont was the first person to recognise gases other than air. He is attributed with coining the term "gas" and discovered that the "wild spirits" (carbon dioxide) produced by burning charcoal as well as by fermenting grape juice, were the same. Because Helmont applied these observations of chemical principles to digestion and nutrition, he is also known as the "Father of Biochemistry". In one of his experiments, Helmont planted a 5 pound willow shoot in a tube containing 200 pounds of dried soil. The tube was covered to keep out dust. After five years of regular watering with rainwater, he discovered that the willow shoot increased in weight by 160 pounds, while the soil lost less than two ounces. Helmont concluded that plants obtain substances for growth from water and this conclusion was correct. However, he failed to realise that they also require carbon dioxide and oxygen from the air. Helmont's collected works, Ortus Medicinae, were published in 1648.

During the centuries that followed, many brilliant minds applied their genius to the task of discovering and defining plant physiology. John Woodward (1665-1728) a fellow of the Royal Society of England, grew plants in water containing various types of soil; you could argue that this was indeed the first man-made hydroponics solution. Woodward found that the greatest growth occurred in water which contained the most soil. However, in those days, little was known about chemistry and he was not able to identify specific growing elements. Woodward concluded that plant growth was a result of certain minerals and substances in the water, derived via the enriched soil rather than purely from water itself. In the decades that followed John Woodward's research, European plant physiologists established many findings. They proved without doubt that water is absorbed via the plant's roots, that the water passes through the plant's stem system and that it escapes into the air through pores in the leaves. They showed that the plant's roots take up minerals from either soil or water, and that leaves draw carbon dioxide from the air. They also demonstrated that the plant's roots also take up oxygen. Later progress in identifying these substances was slow until more sophisticated research, techniques and technology were developed and made available, that further advances were made. The modern theory of chemistry made massive advances during the 17th and 18th centuries and these advances subsequently revolutionised scientific research. Plants, when subjected to analysis, consisted of only elements derived from water, air and soil. Sir Humphrey Davy (1778-1829), inventor of the Safety Lamp, had evolved a technique of effecting chemical decomposition by means of electric current. The experiments highlighted several of the elements which go into the make-up of matter. It was now possible for chemists to split a compound into its constituent parts. The brilliant English scientist Joseph Priestly (1733-1804) discovered that plants placed in a chamber having a high level of "fixed air" will gradually absorb the carbon dioxide and give off oxygen. Jean Ingen-Housz two years later, took John Priestley's work one step further, and demonstrated that plants placed in a chamber filled with carbon dioxide could replace the gas with oxygen within several hours if the chamber was placed in sunlight. However, as sunlight alone had no effect on the container of carbon dioxide, it was certain that the plants were responsible for this remarkable transformation. Ingen-Housz went on to establish that this process worked more quickly in conditions of bright light and the opposite occurred in dim light. He also concluded that only the green parts of the plant were involved. Nicolas Theodore de Saussure (1767-1845) in 1804 proposed and published results of his investigations and findings that plants are composed of mineral and chemical elements obtained via water, air and soil. By 1842, a list of nine elements believed to be essential to plant growth had been identified. These propositions were later verified by Jean Baptiste Joseph Dieudonné Boussingault (1802-1887) who was a French scientist who began as a mineralogist and turned to agricultural chemistry in the early 1850s. In his experiments with inert growing media which consisted of feeding plants with water solutions of various combinations of soil elements and growing the plants in pure sand, quartz and even charcoal which all can be defined as inert growing media i.e. not soil, to which were added solutions of known chemical compositions. Boussingault concluded that water was essential for plant growth in providing hydrogen and that plant dry matter consisted of hydrogen plus carbon and oxygen which came from the air. He also stated that plants contain nitrogen and other mineral elements and that the plants derived all of the nutrient required from soil elements he used. Boussingault was then able to identify the mineral elements and what proportions were necessary to optimise plant growth which was a major breakthrough. These people's work were but a few scientists, who advanced research into the building blocks of the modern hydroponic garden.

In 1856, an industrious scientist named Salm-Horsmar made a great step in the evolution of hydroponics. He was the first to develop techniques for growing plants in sand and other inert media using solely water and nutrients. Using the knowledge gleaned from his predecessors, he saturated the medium with minerals that were known to sustain plant life. The next step was to remove the growing medium completely and grow the plants in just water containing these minerals.

A few years later in 1859 and 1865, the first true hydroponic water culture was formed. The media was completely removed and plants were grown supported by water and nutrients alone. Two German scientists accomplished this feat; Julius Von Sachs (1832-1897), the professor of Botany at the University of Wurzburg I in 1860 and W Knop an agricultural chemist in 1861. Both of the solutions these scientists developed to sustain their plants were composed of various salts. Knop's formula, as published in 1865 is as follows: KNO^3, $Ca(NO3)^2$, $KH2PO^4$, $MgSO^4$, and iron salt (all concentrations in units of grams/litre). This formula and many variations of it, is still in use today. Knop has been called "the father of water culture". Sachs was most renowned for discovering the importance of transpiration in plants and the role of chlorophyll. His in-depth research into plant metabolism was a major contribution to the advancement of hydroponics. Sachs published the first standard formula for nutrient solution that could be dissolved in water and in which plants could be successfully grown. This signified the end of a long search for the source of nutrients vital to all plants.

Thus 'hydroculture' was born. This being the origin of nutriculture and similar techniques are still used today in laboratory studies of plant physiology and plant nutrition. These early investigations and experimentations by these two scientists solidified the fact that normal, even accelerated plant growth can be obtained by immersing the roots of a plant in a nutrient rich water solution. The salts of (N)nitrogen, (P)phosphorous, (S)sulphur, (K)potassium, (C)calcium and (M)magnesium are now defined as macroelements or macronutrients (elements required in relatively large amounts).

With further improvements in laboratory practices and chemistry, scientists discovered that seven more elements were required by plants in comparatively small quantities – the microelements or trace elements. These include (Fe)iron, (Cl)chlorine, (Mn)manganese, (B)boron, (Zn)zinc, (Cu)copper and (Mo)molybdenum.

The inclusion of these macro and micronutrients to water was found to produce a true nutrient solution which would support life, so by 1920, the laboratory preparation of water cultures had all but been standardised and the procedures for their use were well established.

Equipped with this knowledge, researchers developed many diverse basic formulae for the study of plant nutrition guided by the original formulae provided by Sachs and Knop. Some of these researchers were Tollens (1882), Tottingham (1914), Shive (1915), Hoagland (1919), Deutschmann (1932), Trelease (1933), Arnon (1938) and Robbins (1946). Many of the formulae developed by these researchers are still in use in laboratory research on plant nutrition and physiology today.

The practical application of nutriculture did not really evolve until the 1920s when the greenhouse industry was attracted to it by its commercial possibilities. The quality of soil in the greenhouse was subject to pests, loss of structure and productivity, and so had to be changed often therefore, researchers became cognizant of the possibilities of nutriculture instead of conventional soil cultivation.

Up until 1930, soilless growing was confined to the laboratory for plant experiments. Words like nutriculture, chemiculture, and aquaculture were used in the 20s and 30s when discussing soilless culture. Between 1925 and 1935 saw extensive laboratory techniques of nutriculture applied to large-scale crop production.

In the late 20s and early 30s, Dr. William F. Gericke of the University of California broadened his lab work on plant nutrition to large scale commercial applications and in so doing termed the word "hydroponics".

Hydroponics means the technique of cultivating plants without the use of soil and employing an inert medium, for example, gravel, sand, perlite, vermiculite, clay pebbles etc and adding a nutrient solution comprising the essential elements needed by a plant in order to grow and develop. Many hydroponic methods use some type of medium that contains organic material like rockwool or clay pebbles, so it is often called soilless culture, but water culture by itself is genuine hydroponics.

So today, hydroponics is used to describe the different methods of growing plants without soil. These techniques, commonly called soilless gardening, include growing plants in pots filled with water and a number of media which can include rockwool, sand, vermiculite and other more alternative media, such as rubble, fragments of cinder blocks, and even polystyrene.

There are several good reasons for using a sterile medium instead of soil. Soil-borne pests and diseases are instantly done away with and the work of looking after your plants is greatly reduced, such as weeding.

What's even better is that you are able to raise plants in a soilless medium that lets you grow more plants in less space, crops mature more quickly and yields are much greater. Hydroponics is typically recirculating, so plants only uptake the water and nutrients they need. It allows you far better control over your plants and guarantees more uniform and consistent results. These results are attainable because of the connection between the plant and its medium as the plant doesn't need soil per se; it's the water and the nutrients in the soil that are necessary to sustain the plant, and the support that soil provides. However, any medium will provide this support. Since the plants are grown in a sterile medium, exact amounts of nutrients and water can be administered to the plants when required. As soil absorbs water and nutrients away from plants, it makes the administration of nutrient in precise doses very difficult indeed, however, in hydroponics, nutrients are dissolved in the water and administered at set times.

Until 1936, this method of cultivation was confined to the laboratory in order to research plant and root development.

Dr. Gericke grew many types of vegetables using hydroponics which included radishes, carrots, potatoes, and cereal crops, fruits and flowers. He employed the water culture technique in large reservoirs and succeeded in growing tomatoes up to a height of 25 feet as evidenced by photographs of him on a step ladder. These photos appeared in the media all over the US, and although quite sensational at the time, this system was ahead of its time for commercial use. It was rather fragile and required round the clock supervision of the equipment.

Budding hydroponicists found the Gericke system

too technical and difficult to construct. Gericke's system was made up of a series of troughs on which wire mesh was overlaid and onto which was spread another suitable mulch over this mesh. The plants were positioned on this mesh in such a way with the roots hanging down through the mesh and into a water and nutrient solution in the troughs.

The problem with utilising this method was getting enough oxygen in the nutrient solution. The plants would quickly use up the oxygen by taking it up through their roots, and so it was extremely important that a constant supply of fresh oxygen was injected into the solution via aeration. Supporting the plants was also tricky in order that the growing tips of the roots were suspended in the solution correctly, and at all times.

The American Press, as ever, made many absurd declarations and hailed it as the discovery of the century. There followed a bout of shady salesmen who tried to cash in by selling substandard equipment and materials that often didn't work. However, more practical research was done and hydroponics became established on solid scientific research in horticulture, especially with regards to high crop yield and its use in infertile regions of the world.

In 1936, W. F. Gericke and J. R. Travernetti published an account of the successful results of cultivating tomatoes in water and nutrient after which several commercial growers started testing some of the techniques, and a few agricultural colleges began working to simplify and refine the process. Several hydroponic units, and some on a huge scale, have been built in Mexico, Puerto Rico, Hawaii, Israel, Japan, India and Europe.

Dr. Gericke's application of hydroponics came into its own when it supplied food to the troops stationed in the Pacific islands during the early 40s. The first real achievement in hydroponics was when Pan Am Airways decided to build a hydroponicum on Wake Island in the Pacific Ocean so as to provide the passengers and crews of the airlines with fresh vegetables. At this point, the British Ministry of Agriculture took a keen interest in hydroponics and its importance in the Grow-More-Food Campaign during the war (1939-1945) was fully appreciated.

In the late 1940s, Robert and Alice Withrow of Purdue University originated a more viable hydroponic method. Inert gravel was utilised as the medium and by flooding and draining the gravel in a container in sequence, maximum amounts of both nutrient solution and air was made available to the roots. This became known as the gravel method of hydroponics and is sometimes also called nutriculture.

During the war, shipping fresh vegetables overseas to military outposts was not feasible, however, hydroponics solved the problem. During World War II, using the gravel method, hydroponics underwent its first real trial as a method of providing fresh vegetables by the US Armed Forces for its personnel. One of the first of many large-scale hydroponics farms was built on Ascension Island in the South Atlantic. This completely barren island was a fuel stop for the US Air Force. A large force was based there to service planes, so all food had to be flown or shipped in. Fresh vegetables were essential, and so a hydroponics installation was built by the armed forces. The methods developed on Ascension Island were later replicated on various islands in the Pacific.

Wake Island, an atoll to the west of Hawaii is not usually able to produce crops, as the rocky terrain precluded traditional farming. The US Air Force built small hydroponic growing beds that consisted of only 120 sq ft of growing area. However, once in production, it was yielding 30 pounds of tomatoes, 20 pounds of string beans, 40 pounds of sweetcorn and 20 heads of lettuce a week! The US Army also constructed hydroponic growing beds on Iwo Jima using crushed volcanic rock as the growing medium and had equivalent yields.

At about the same time, the Air Ministry in London initiated soilless culture at their desert base of Habbaniya in Iraq, and on the island of Bahrain where vital oil fields were located. At Habbaniya, all fresh produce had had to be flown in from Palestine to feed the troops stationed there, and this was a costly endeavour.

Millions of tons of hydroponically grown vegetables were consumed by the Allied Forces during the war years and after World War II, the military continued to use hydroponics. The US Army had a dedicated hydroponics department which grew over 8,000,000 lbs of fresh produce in 1952. The US Army was also responsible for building one of the world's largest hydroponic installations of 22 hectares in Chofu, Japan. Hydroponics in Japan was essential to the troops based there because the Japanese fertilised their soil using night soil, which contained in the main, human sewage. It was highly contaminated by bacteria to which the Japanese were immune, but the occupying forces were not. It was operational for over 15 years and this site, as well as others, have been extremely successful.

After 1945, several commercial installations were set up in the US, notably in Florida. The majority were outdoor set-ups and subject to weather conditions. Substandard building methods and bad management caused many of them to fold or produce inconsistent yields. Still, the commercial use of hydroponics expanded throughout the world in the 50s in countries such as Italy, Spain, France, England, Germany, Sweden, the former Soviet Union and Israel, to name but a few.

The early pioneers of hydroponics were faced with many drawbacks and one of these was the concrete used for the growing beds. Lime and other elements percolated into the nutrient solution plus the various elements in the solution had an impact on the majority of the different metals used. Lots of these early hydroponic installations used mainly galvanized and iron piping and they corroded rapidly, but in addition, harmful and toxic components were released into the nutrient solution and that had adverse effects on the plants.

However, enthusiasm remained high in hydroponics due to a number of factors. Firstly, no soil was needed and a large number of plants could be grown in a small area. Secondly, if correct amounts of nutrient were administered, then high yields of crops would be realised. Hydroponically grown vegetables as opposed to conventionally grown ones were generally of a better quality, and growth was accelerated and enhanced. Also, hydroponically grown vegetables had a longer shelf life.

A point of interest was that many oil and mining companies constructed large hydroponic installations at some of their facilities in various parts of the world where traditional farming techniques were not viable. Some were in the desert where there was little water available. Other facilities were on islands where there was little or no soil available for growing vegetables.

Big commercial American businesses had many acres set aside for the cultivation of vegetables for city dwellers in the Far East, while various oil companies in the West Indies, the Middle East and the Sahara Desert and Venezuela have found hydroponics crucial to guaranteeing that their employees got fresh vegetables.

In the US, substantial commercial hydroponic farms exist, which produce vast quotas of food on a daily basis, particularly in Illinois, Ohio, California, Arizona, Indiana, Missouri and Florida and there has been some development of note in hydroponics in Mexico and parts of Central America.

Apart from aspects of hydroponics that had focussed on the commercial systems built between 1945 and the 1960s, a lot of research was done on the smaller applications of hydroponics for apartments, houses and in the garden, for growing both flowers and vegetables. Much of this was not successful for several reasons: inferior media, inappropriate materials particularly in the manufacture of troughs that were used as growing beds, and rudimentary environmental controls.

Despite these setbacks however, hydroponic growers everywhere were sure that these minor problems could be overcome. It was also felt that due to diminished food output and the increasing world population, hydroponics was the perfect method and answer to these problems.

With the USA and the UK already being established hydroponics enthusiasts, Russia, France, Canada, South Africa, Holland, Japan, Australia and Germany also became hydroponically enlightened. Apart from producing vegetables however, between 1930 and 1960, similar research was being done for the production of livestock and poultry feed. Researchers had discovered that by growing cereal hydroponically, turnaround of yields was greatly accelerated. Grains such as barley were used and they showed that 5 pounds of seed could be converted into 35 pounds of abundant green feed in as little as a week. As a supplement to normal quotas of food, this was very beneficial to many varieties of animals and birds. Milk flow was increased in lactating livestock and when breeding stock, the fertility of males and conception in females rose noticeably. Poultry also fared well in a number of ways in that egg production increased whereas cannibalism, which had been a persistent problem for poultry farmers, had stopped.

However, there were also problems of consistent yields, and these early systems had few or no temperature or humidity controls, so growth rates were varied. Mould and fungi were constant problems. Clean seed grains with a high germination ratio were essential if good growth rates were to be realised.

Despite these problems, a few committed researchers carried on working to refine a system that would be able to supply uninterrupted quotas of this crop. With the onset of new techniques, equipment and materials, low maintenance and easy to use hydroponic units were becoming increasingly available.

Hydroponics did not get to India until 1946. Here, from the very start, a number of problems particular to this country became apparent. A brief study of the various commercial methods in use in Britain and America demonstrated how impractical it would be for use by the Indian population. A fresh approach, which was both practical and cheap, would have to be at the core if hydroponics was to succeed in India, and indeed in other parts of Asia.

One objective was to remove the complex mechanisms of hydroponics and offer it to the people of India and the rest of the world, as an inexpensive and simple way of growing vegetables without soil. Today, thousands of homes in India grow vegetables in user-friendly set-ups on rooftops or in gardens.

All over the world, the benefits of hydroponics have been felt. But judging by the accounts in the media, they would have us believe that hydroponics is a new technology which will solve the world's food problems. It shows little research on the part of the media. Undoubtedly it will help with food shortages, but as to its being a new development is another matter. As it is, the first plants on the earth were grown hydroponically and more than 50% of plant life is growing hydroponically in nature. In fact, plants growing in the oceans, which cover more than 70% of this planet, are the healthiest examples of hydroponics in nature as here there is no soil in the oceans, so plants get all their nutrients directly from Nature's most complete hydroponic nutrient solution available, which is sea water.

Contributors to the establishment of hydroponics are the Universities of Illinois, Ohio, Purdue and California in the US, Canada's Central Experimental Farm at Ottawa, the University of Reading in the UK which is renowned for its groundbreaking research in new cropping techniques, as well as ICI Ltd which adapted hydroponics to British conditions, and Dr Allen Cooper who invented the Nutrient Film Technique and refined its practical application via his work at the Glasshouse Crops Research Institute, now called HRI Littlehampton in the UK. The establishment of commercial NFT in cropping enterprises the world over has been attributed to the work of this visionary man and even today, the full extent of Dr Cooper's innovation has not been fully realised. Other pioneers of hydroponics were the New Jersey Agriculture Experiment Station, USA; the Boyce Thompson Institute for Plant Research, New York, USA; the Horticultural Experiment Station, Naaldwijk in the Netherlands and the Alabama Polytechnic Institute, USA.

The Present

Another great leap forward in hydroponics was the evolution of plastic which can be credited with revolutionising the industry. The leaching of harmful elements into the nutrient solution from concrete, metal, rooting media and other materials was a problem. Fiberglass and plastic such as vinyl, polythene film and the many kinds of plastic pipe, practically eradicated these problems. Superior kits that are constructed today use plastic throughout and there is little or no metal used and even the pumps are plastic coated. With this type of material and using an inert medium, success is within the grower's reach. Plastics liberated growers from the costs of building concrete beds and tanks. With the advent of pumps, timers, plastic tubing, solenoid valves and other technology, the hydroponic set-up could be fully automated and even computerised, thus minimising initial set-up and running costs.

One of the mottoes of hydroponics is keeping it simple. As with most new inventions, it all seems confusing and complicated at first, which is how it is with hydroponics. But once you get your head around the concept, you'll see how simple it really is.

In addition to plastic and in improved nutrients, was the refinement of environmental controls which had a huge impact on the industry. The improvements in hydroponic systems has been responsible for the huge growth in hydroponics in the last 50 or so years, and there can be no doubt that hydroponics will be a big player in future food production.

In hydroponics, soil is not a requirement and only about $^{1}/_{25}$th as much water is needed as compared to traditional farming methods. The future may well show installations on the rooftops of warehouses or other large buildings. One such example, although it is no longer in existence, was the Deutschmann's Hydroponic Centers of St. Louis. At its peak, there were 7 rooftops for vegetable production. Unfortunately, the company suspended operations, although unrelated to the business, but the patented planting system developed by the inventor is still available today. The system designed and built in St. Louis shows that the technology to build such systems relatively cheaply, is already available. However, systems built by or for the amateur grower that takes up little space and which are small enough to be put in lofts or basements or other parts of the house and can grow an abundant array of food such as lettuce, strawberries and herbs, are available now. Any flat rooftop is suitable for hydroponics. All you need, apart from the space, are electricity and water and because the environment can be controlled, vegetables can be grown all year round in practically any climate.

The Canary Islands has devoted hundreds of acres of land with polytunnels that house tomatoes that are grown hydroponically. These structures are positioned in such a way so that the wind blows through and cools the plants. The polytunnels also reduce water loss from the plants and protect them from storms. Just about every state in the US has substantial hydroponic installations and Canada also grows hydroponic vegetables in greenhouses on an extensive scale. Fifty percent of Vancouver Island's tomatoes are hydroponically produced and one-fifth of Moscow's tomatoes are also hydroponically produced. There are fully working hydroponic set ups on American Nuclear Submarines, Russian Space Stations, and on off-shore drilling rigs. Every NASA space shuttle flight has had a hydroponics experiment incorporated into its missions.

Hydroponics is a fact of life for growers in practically all types of climates. Huge installations can be found all over the world, such as hydroponic greenhouses in Arizona, USA (about 15 acres) and Abu Dhabi in the Middle East which is over 25 acres. This particular installation uses a desalination facility to water its crops. In countries such as Mexico and the Middle East, where the supplies of fresh water are in short supply, hydroponic installations incorporating desalination units are being developed. These are near the ocean and the medium used is the existing beach sand. In the Middle East for example, land suitable for farming is in short supply, but because of the oil industry and the ensuing wealth, large hydroponic installations farms to feed the increasing population in these countries was unavoidable.

The Future

Hydroponics is still in its infancy. It has been used on a commercial basis for several decades but even in this short period of time, it has been utilised in so many different ways such as farming, greenhouse horticulture and even in specialist fields such as military submarines and Russian space stations This is the space age on the one hand, and on the other, can also be used by Third World countries for intensive farming in a small area. Its only limitations are fresh water and nutrients and where fresh water is not easily accessible, desalination is an option if close to the sea. So, there is excellent potential for supplying food in regions having large areas of non-arable land, such as deserts. Hydroponic installations can be established in areas of land that cannot be cultivated and in combination with desalination plants while using natural resources such as sand as the medium for growing the plants.

Hydroponics is an excellent way of growing fresh vegetables in countries that have to support large populations, but have little land available that is agriculturally viable. It would certainly benefit small countries whose main industry, for example, is tourism where hotels have usurped the vast majority of farming land and forced local agriculture out. In this scenario, hydroponics could be located on non-arable land so as to supply fresh vegetables for local and tourist populations. For example, tomatoes grown hydroponically could yield 150 tons per acre per year, so a 10 acre site could yield 3 million pounds of tomatoes a year.

However, the development of hydroponics continues to be hindered, particularly by defeatist attitudes of people in authority, especially in government and educational establishments. These viewpoints range from a complete lack of interest to antipathy, usually because they have failed to obtain the yields achieved by other growers. There is hope though, as some enlightened people within the educational establishments have devoted their time and expertise in this area.

Recently, the costs of energy have risen quite dramatically and this has caused several profitable facilities to start operating at a loss, and some businesses have had to shut down completely in the winter, and since this is when vegetables are at or near optimum prices, the increased prices of fuel have had devastating consequences in this field. The light at the end of the tunnel has been solar heating systems. There has been a lot of research in this area and there are many off-the-shelf systems available on the market. Also, there are many publications with plans on how to build your own system. Many new developments in this area are on the horizon and solar energy may be the answer to this fuel price problem.

But what has been the biggest threat to hydroponics has been the invasion of so-called experts in the last 50 years. Hydroponics has been the victim of its own success, and this in turn has created self-appointed experts in ever increasing numbers. Second-rate systems have been peddled on the unsuspecting and ill-informed public. The cost for a commercial hydroponic system, with good environmental controls, can be very expensive, so a reputable hydroponics shop should be your first stop.

Anyone who is willing to commit themselves, and persevere, can grow hydroponically – even the novice. Anyone can produce vegetables all year round and enough for a family of four or more. As with any successful business, hydroponics can be profitable, if the time and effort is put into it. Hydroponically grown tomato yields can be 18 times that of those grown in soil, and pesticides need only be used in the most extreme cases, especially with the biological controls available today. Hydroponics can bring many hours of joy to the amateur or expert grower alike.

Hydroponic techniques are always changing and while new inventions and improvements come along, the basic principle remains the same. Administering nutrient needs moving water and all the essential minerals and elements need to be made available to the plant in a manner that is easily absorbed. Since the era of the Hanging Gardens of Babylon, hydroponics has significantly evolved as an agricultural method. Climate changes, population increases and deforestation have played their part in the dwindling land that can be farmed for crops. Japan, Mexico, India and Hawaii are four regions that are faced with this problem. This could be due to an increase in tourism, over population or simply soil erosion – whatever the case, traditional farming methods are not able to support the local and visiting populations. Hydroponics is fast becoming the preferred method of cultivation in these areas because of the rapid growth times and maximising the limited space available. Both commerce and government are benefiting from hydroponics, and profiting financially. Need is a great motivator and this need has propelled hydroponics, and as we move through the 21st century, hydroponics is becoming the obvious method of feeding the world.

Hydroponics has been used and modified by commerce in order to acquire speedy results in their enterprises. Aquaculture and nutriculture have forged a relationship in the world of agriculture. What was once the reserve of ecospheres has now been embraced by farm fisheries and hydroponic crop producers who are constructing combined facilities to support and maintain both their activities. NASA is forever improving hydroponic techniques in order to provide food for space exploration and colonisation. Specialty growers are supplying the restaurant industry with fresh herbs and salad greens which are available all year round. The applications for this science are almost too numerous to list.

This technology allows you to grow where no one has grown before, be it underground, or above, in space or under the oceans, this technology will allow humanity to live where humanity chooses. If used for our own survival or our colonisation, hydroponics is and will be a major part of our collective future.

Appendix 1
Transcript

What follows next is a transcript taken from a presentation and demonstration on how to get started and how to maintain a particular hydroponics system. This information can be adapted and applied to any hydroponics system. It is advisable to read the whole transcript to get a basic overview of the complete process.

Guide to Propagation and Nutrient Solutions

Presented by Jeffrey Winterborne

We're here today to talk about jiffy pellets and rockwool for seed germination. [picking up rockwool and jiffy pellet] On my right (probably your left), jiffy pellets, compressed peat pellets, completely dry and on my left (probably your right), rockwool. Now rockwool is inert and sterile which means we have to supplement it with nutrient and also pH-adjusted water to compensate its own consistency. Now jiffy pellets, on the other hand, has got everything you could require for seed germination or for even taking cuttings, apart from the essential part which is obviously water. There's enough nutrient in there, there's enough compost in there to keep a cutting or a seedling going until it's got a good healthy root system. Jiffy pellets are ideal for beginners, completely and utterly due to the fact that all you have to do, [drops jiffy pellet into container of water] is soak them in water.

Now the key to jiffy pellets is only to soak them for approximately five minutes. Now that is dependent on how hot or how cold the water is; if it's very warm water, it'll take less time, if it's very cold water it will take more time. Now what we'll do is that as soon as that jiffy pellet has swollen to its potential, I'll pull it out to show you exactly how swollen it should be. The key with jiffy pellets is DO NOT oversoak them, I cannot stress that more. If you oversoak a jiffy pellet, they end up waterlogged.

When they're waterlogged, if you put a seedling in, the seedling will germinate, or the cutting will take, but because they're so waterlogged, no air to the root system and the root system will starve resulting in possibly a dead plant. If not a dead plant, a very stressed plant.

Now we'll let the jiffy go in the background while I talk you through rockwool. Now this is the more professional way to cultivate seedlings or cuttings, but it takes a few variables to iron out before you get it absolutely right. Now rockwool on its own with plain water wouldn't be good enough, you have to supplement it with nutrient and then you have to pH-adjust the water and the nutrient solution to compensate its own consistency, which we're going to do right now. Now, an ideal nutrient to use for seedlings or cuttings [picks up bottle of Formulex] is one called Formulex. Now Formulex is superb; it's tailor-made for rockwool cultivation; however any Grow A and B nutrient will suffice as long as you make it up to a weak CF level. Now CF, TDS, PPM; they're all measurements to verify the electrical conductivity of the nutrients in the water, in layman's terms it means how strong the food is in the water.

How strong your nutrient is...now Formulex is designed already to be very low in the EC value, ok, it comes with full instructions on the back and tells you the dilution ratio. Now you can follow that to the 'T' and you'll end up in the right ballpark; what we're going to use, [picks up meters] being more professional, are meters. Now, we have a pH meter [holds meter up] and we have a CF meter [holds meter up], also called a TDS meter, also called a PPM meter. It's the same thing; it just measures how strong the nutrient is in the solution. I've just poured approximately 4 litres of water into this container for you to be able to see. Now what we've basically got to do is raise or add enough Formulex into the solution to bring the CF level up to approximately 12. Anywhere between 10 and 12 will suffice. But what is a good practice to do [picks up CF meter, opens box, takes meter out] before adding any nutrient to your stock solution is just to test before you do so, [puts meter into container of water] the conductivity factor of the water without any nutrients in it at all. Now it's given me a reading of 4, that's a TDS level of 4. [takes meter out of container of water] What that means is, there's basically a conductivity factor of 4 which is dissolved salts into the water already, but that is what they would call blind water or blind salts.

The salts they've gathered are actually dead salts. So that's given us a reading of 4, well that 4 is actually misleading us because it has no usable nutrient in there because it's dead already, ok, so we have to make a note that that's reading 4 and add that to the equation. Now 4 is quite high, so we always recommend halving the background nutrient reading and adding that to your stock solution. So if you're looking for a CF of 12 for the background for the rockwool, [picks up rockwool cube] you want to add another 2 to the equation because we've already got 4 in the water but that's 4 of dead salts, ok.

Now, to put that in English, we've got food in the water which is unusable food and what we're going to do is add food to the water. Now this pen reads exactly how much unusable salts are in the water. So what we've got to do is take that and compensate against how much of this we put in. In so doing, it's reading 4, we're going to say ok it's misleading us by 2, because we're halving the original amount just to make sure that we're in the right ballpark. If you over feed in hydroponics you end up coming unstuck, if you under feed, you always have a result, so always do everything by half and then look to step up. Now, what we're going to do [picks up bottle, shakes it and opens it] is add some Formulex into the stock solution [picks up meter in the other hand] to raise the water up to a CF level of 12. Twelve is absolutely ideal level for seedlings and cuttings, ok; it's been tried and tested and scientifically proven that a level between 10 and 12 is approximately what a seedling or cutting would require with a little bit excess just in case it does have a lot of vigour to grow. Anything above 12, you'll be doing your seedlings a lot of damage because it will be too much dissolved salts in the water for the seedling to absorb. Now that would cause a problem to the rootball; it could burn it and therefore cripple your plant.

Now, we've done this quite a few times, so what I'm going to do is put a wee splash into the solution, [adds solution to water] ok, now I'm going to give it a little stir, [stirs solution with meter] it's not the ideal thing to stir with, is a nutrient meter, but it will suffice and then take another reading. Ok, we're up to a CF reading of 6 now, so we've got to add some more. Now I'm just going to get some pipettes to do it properly [walks off set to get pipettes] Ok, pipettes are well worth using obviously because they have millilitres measured on the side of them. Now, in this, it basically explains that in hydroponics you want one capful to 1.5 litres of water, ok, now instructions are great but it's always best to do it yourself and confirm it with the meters; you know then that you've got it approximately spot on. [dips pipette into bottle] So what we're going to do is suck up another 4 ml, splash that in and give it a stir. And it's always worth bringing it up gradually, slowly and that way you don't have to [puts meter into solution] keep pouring nutrient away in order to get it right. We're now up to a CF level of 8, ok, so approximately another 3 ml [puts pipette into bottle, adds to the solution, stirs it] should bring it where we roughly need it. [puts meter into solution] Ok. We're now up to CF level of 12, now 12 in this case is perfect, 10 to 12 is really where we want it.

Now, once we've added nutrient, we then test for pH. [picks up pH meter, takes it out of the case] You always test pH after CF because adding nutrient to your stock solution adjusts the pH naturally. Now, one pH meter, popped in [puts meter into solution] is telling us that we have a pH reading of 7.3, now 7.3 is very high, too high for rockwool. Ok if you soak that in that solution, put the seedling in, the seedling will be stressed because the pH is far too high for it to grow happily.

Now what we have are 2 solutions, [picks up a bottle in each hand] phosphoric acid, ok, and potassium acid. Now potassium acid is used to raise pH in your stock solution where phosphoric acid is used to lower pH in your stock solution. Now the reason that these 2 bottles are in bags, sealed, is because when mixed together in a concentrate, they're diametrically opposed which means that they conflict to such a degree that they could explode and if these explode in your face you end up in the hospital explaining why you've got potassium and phosphoric acid all over you. Whatever you do, always keep these bottles separate; always administrate them into your reservoir separately. What we recommend doing is actually having one pipette per job [picks up pipette, one in each hand with the bottles], get some elastic bands, get some ink pens and actually colour code one pipette per bottle. That way you'll never get them muddled up. If you end up taking a pipette full of pH down, putting it in your tank and then putting that pipette into some pH up, the pipette will explode and it's not a very nice scene. BE WARNED!! Never mix the 2 together; they're quite safe diluted in a solution, but as a concentrate, they're very, very aggressive.

Ok, now what we're going to do is use pH down [puts pH up bottle and pipette down on left hand side] because in this case, the pH level is too high. So, we'll crank it out of the bag, [removes packaging from bottle] one pipette over there [puts one pipette on left hand side] get rid of that [throws packaging down], give it a little shake. [shakes bottle, removes cap] Now, pH down is very, very aggressive, so you only have to administer tiny amounts; in this case, [dips pipette into bottle] all we are going to administer or administrate [puts pH down bottle down on right hand side] I should say, is the amount on the exterior side of the pipette. [puts pipette into solution and stirs]

Ok, so we've not actually put any in apart from the residue that's on the outer edge of the pipette. [puts pipette down on right hand side, and picks up meter] Then, we'll stir it up and we'll retest. We could have possibly put 1 or 2 drops in but as I said earlier, it's always best to underdo everything, do everything slowly and do it again and again and you'll get it roughly where you need it.

Ok, we've just lowered the pH through doing that down to 7.2, so what we'll do is we'll do that again, [picks up bottle, dips pipette in] but in this case add a little drop as well, so one tiny drop [puts pipette into solution and stirs] and a little bit of exterior on the pipe, pop it in, give it another stir, [picks up pipette and stirs again, puts the meter back in] ok, we're down to 6.9 so we've still got a little way to go, [picks up bottle of pH down again, puts pipette into the bottle] one more little drop, [puts pipette into solution, stirs, puts bottle and pipette down on right hand side, puts meter in again] Ok. We got it down now to 6.1, 6.1 so one more go at that, [dips pipette into bottle, puts pipette into solution, gives it another stir] give it a good stir, [puts meter in again] ok, 5.7, 5.6... 5.6. Right, that's good enough, the value I was looking for was 5.5 , but 5.6, 5.5 is roughly where you need it. Good enough, you don't have to get it absolutely spot on, if you take it too far down then you've got to use pH up which you don't want to do, you'd have to throw the stock solution away and start again. You don't want to use a lot of pH up and pH down together at this stage for seedlings or cuttings. If you use too much of it, it's very, very aggressive and it can do the root system a lot of damage. [picks up rockwool cube] If you end up taking the pH too far down, you're much better off throwing the stock solution away and starting again from scratch. Ok so, we've the level now to 5.6, which is good enough. Ideally, I would have wanted 5.5, but 5.6 will suffice, I don't want to take it down any further in case it goes too far down.

Now, what you've done basically is added nutrient into the solution to a level of CF of 12 or TDS of 12, we've then adjusted the stock solution down to a pH level of 5.6; we're now ready to soak the rockwool cube. [drops rockwool cube into solution] Now, while that's soaking; [removes jiffy pellet from solution] our jiffy pellet is finished ok. What was a very squashed peat pellet, [picks up dry peat pellet in other hand] compressed, is now a very swollen sack of dirt. Now that is basically the consistency where you'll want it where if you give it a slight squeeze, [squeezes jiffy pellet slightly] it drips. It's wet enough to keep your seedling going for at least 3 or 4 days, but it's also wet enough to let air in for the root system to breathe. Now what you would do with a jiffy pellet; I'll use a different pipette, [picks up a pipette and pokes the pellet] is make a hole approximately 3 mm beneath the surface, 3-5 mm; you'd then pop your desired seed into the jiffy pellet and then cover the seed over, ok, you'd cover it over to a degree where it is cocooned in the dirt itself.

The seedling has to be cocooned to keep the moisture completely around it to encourage the seed to germinate. At that point, it is then ready to go into your propagator. Perfect. [puts jiffy pellet down] Now, while that's been soaking, [puts hand into container and removes soaked rockwool cube] our rockwool has soaked down to the bottom, now some people like to soak rockwool overnight; it allows the chlorine to rise and that way, when you pull the rockwool out it doesn't have a lot of chlorine in the actual rockwool itself. However, you don't have to, I mean to be quite frank, as long as its soaked and sunk all the way down to the bottom, it's good enough to receive the seedling. Now the key with rockwool is now to [squeezes rockwool cube slightly] give it a squeeze, ok, the idea of a squeeze is to get approximately 5-10 % of the volume of water out of the rockwool in order to get oxygen in. If we put a seedling in without the squeeze, you end up with a waterlogged rockwool block. When you put the seed into a waterlogged rockwool block, same thing like the jiffy, it ends up rotting because it cannot breathe properly.

A seed or a cutting needs just the right amount of food, not too much, just the right pH level and just the right wetness. Not too wet, but more wet than dry, so the key to it is just to squeeze 5-10% of volume out of the rockwool itself. So we've got water in there, we've got nutrient in there, we've got air in there; we've got everything the seedling needs. We'll then make, [picks up pipette and pokes hole into rockwool cube] just like in the jiffy pellet, approximately a hole 5 mm beneath the surface, we then get the seed and pop the seed into the hole and then gently cover the seed over. Now with rockwool, do not pinch it tight. When you're covering the seed over, you have to do it gently. A seedling only has enough inertial energy to get up ok, if it's blocked, if this rockwool has been squashed too tightly, it will not fight its way through to the surface, it'll end up dying. So you just cover it over gently and then it's ready to go into your propagator. That's how you do rockwool, that's how you do jiffy pellets.

pH meters are semi-waterproof [dips meter into container] up until the first ridge on the meter. If you accidentally drop them in, fully submerge them, you want to pull them out and turn them off and put them on a radiator and dry them out. There's a good chance they'll come back to life, however, they might be completely and utterly fried. So they are semi-waterproof but not fully waterproof and they work after; well it's a good idea to give them about 5 seconds in order to give a proper reading because it can fluctuate because of the temperature in the water it has to compensate for. That's a pH meter, and then always turn it off after use [puts pH meter down and picks up CF meter] because if you leave it on after use, it ends up wearing the batteries out which is going to give you a fake reading the next time you use it. So switch it on, [dips CF meter into container] dip it in, get the reading you want, turn it off. At that point, you then adjust your solution.

OK, we've just talked about pH meters and CF meters. [Takes a meter in either hand] pH [puts right hand up with pH meter] CF [puts left hand up with CF meter] also known as TDS, PPM, so on and so forth. It measures the amount of food in the reservoir, [lifts left hand up and forward] that measures the pH level in the reservoir. [lifts right hand up and forward]

Now it's very good practice once every couple of weeks to calibrate your meters. If you don't calibrate them, you don't know that they're reading a true reading. In so doing you could be feeding your plants too much or too less or you could be getting the pH completely and utterly wrong.

Now, on a pH meter, [keeps meter in right hand and picks up a bottle with left hand] you use a solution called Buffer 7. Buffer 7 is specifically balanced at a pH level of 7, ok what you'd basically do with this is pour it in a cup, get the meter in and then see exactly what the meter is reading. There's a little screw on the back [turns meter over and points to back] and a little screwdriver that comes in the case [takes screwdriver out of case] and that is how you would adjust the level of calibration.

What we're going to do is actually do it right now [puts meter down, removes cap from bottle] so that you can see. So what I'm doing is pouring Buffer 7 into a glass, [pours solution into glass] we then turn the meter on and then dip it in. Ok. Now what we've got is a reading of 7.2, so 7.2 is slightly high [picks up screwdriver and adjusts meter] so it needs a little bit of adjustment. So we put the little screwdriver in the back, give it a little twist one way – I've gone too far that way so then I go back a little bit the other way – a little bit of fine adjustment back and forth to get to a pH level of 7. [removes meter from glass] Now that meter is now perfectly calibrated pH 7 so we know it's true. That's how you pH calibrate.

This is how you CF/TDS calibrate. [picks up CF meter in right hand] This pen uses a different calibration solution. [picks up bottle in left hand] They call it conductivity standard. Basically, it is stable at CF level of 27. So when you pour that into a cup and pop the meter in, the reading that you want to be looking for is 27. If it's too high, if it's too low, same story, put the pen in, the

screwdriver in, twist it one way or the other till you get to 27. We'll do that for you right now just to show you how easy it is. [puts meter down, removes cap from bottle, pours some solution into glass] So pour a little bit into a glass, get your TDS meter, pop it in, and this is actually showing me a reading of 27 so I don't need to do any adjustment, but if you did, just stick the screwdriver in the back and turn it one way or the other to compensate and that is basically how you simply calibrate meters. Now a lot of people think that that's very difficult but I can't see why. Ideally, once you've used your solution in a cup, they recommend to throw it away because the more air and the more contact it has with other surfaces, the more unstable it becomes. However, it's a bit puritanical and you can get away with pouring it back into the bottle. It's good practice to throw it away, but you can get away with not, ok. Which a lot of these fluid manufacturers won't tell you about. That's how you calibrate.

You put the seed into the rockwool or the jiffy pellet [holding jiffy pellet in one hand and rockwool cube in other hand] or even both depending on what you fancy doing at the time, they then would like to go into a propagator. [holds both jiffy pellet and rockwool cube in one hand and picks up the propagator in the other hand] The propagator basically acts as like an external womb for the rockwool or the jiffy pellet. It creates maximum humidity in the environment. In so doing, it stabilises the environment, encourages the seed to germinate. Now we recommend [puts propagator down, removes lid] putting your seedlings and cuttings in – [puts jiffy pellet and rockwool cube into the propagator] they don't want to be sat in a puddle of water, ok, but a little bit of excess water in the drain [picks up propagator to show drain] is not going to hurt, but as I said do not let the rockwool have direct contact with that puddle of water. In so doing that rockwool or jiffy pellet will become waterlogged and that will be no good for your seedlings.

Now what would happen is that you would put them in, you'd shut the lid and you'd then close the air vents and before long the whole chamber becomes covered in moisture, in condensation. You would then put that in a very warm sort of environment – airing cupboard. If you have a heated propagator you don't have to worry about that, you just turn the heat on, however, that's dependant on the time of year. If it's anything like the weather at the moment which is approximately 30 degrees, you don't need to put that in an airing cupboard, ok, it'll do quite happily on its own in a normal room. If it's in the middle of the winter, then ideally it would want to go somewhere like an airing cupboard, just for the initial germination process.

Ok. No light is needed at this point, it just needs to go somewhere warm, warm but not very hot. Now over the course of possibly 24 hours, but it could also take up to 7 days, the seedling will germinate. It will start to grow [puts propagator down and removes lid] and it will start to develop. Now initially, [removes rockwool cube from propagator] it will, basically, it won't look much like a plant at all. As soon as you see any sign of life, anything breaking through the top of the rockwool or through the top of the jiffy pellet, [picks up jiffy pellet] then it's time to employ some form of lighting. Either put that by the window sill or put actually under your light. Now, if you put this by the window sill, seedlings cannot tolerate direct sunlight; if they get direct sunlight on a hot summer's day, then there's a good chance that they'll fall over because they won't be able to take the intensity of the light or the heat generated from the sun. So ideally, they would want to go under a propagation light, high frequency fluorescent lights are perfect for that, mercury blend 250 watts are perfect for that.

Now if you are going to use an industrial light, like the high density discharge, metal halides or high pressure sodiums, you want to make sure they're an adequate distance away from the canopy of the propagator. If you are using a 400 watt high pressure sodium for example, I would want it approximately 3-4 foot away from the canopy of the propagator. That should have enough distance between the light to the actual seedling itself that the light dissipates down to such a degree that it's adequate light for them, but not too intense.

Right, with a 400 watt light, approximately 3-4 foot away from the lid of the propagator, ok, and you would want to run that light, well, I'd have it at least 18 hours a day, possibly 20 hours a day, so about 4 foot away, approximately there from the lid [shows distance between propagator and light] Now I wouldn't run the light 24 hours a day as a lot of grow books say. What you'll find is that if you run a light 24 hours a day for seedlings is that you end up stressing them out and undue stress is not what they need at this particular point in their life. They're very, very fragile and they need an easy breaking in not a hard one. So give them a sleep, the whole of nature to my knowledge requires sleep, why are seedlings going to be any different? These books that state 24 hour lighting, it can work in a lot of cases but it also has a lot of detrimental effects that they don't tell you about. So lighting on for approximately 18-20 hours a day, approximately 4 foot away from the canopy, the seedling will start to grow. Before long, it will develop leaves. [picks up rockwool cube from propagator] Now ideally, what will happen is before it's developed enough leaves, or sustained itself to a degree where it can be transplanted into another system, you would want to break it in to a bigger rockwool cube, i.e. what will happen is roots will grow out of the side and the bottom. They're going to become root bound before you can actually get them into another system. However, they're still very young and still very fragile so what we recommend doing is that as soon as any roots start growing through the side and through the bottom, you then need to transplant that rockwool cube to a bigger rockwool cube and then put it back into the propagator and propagate for some more. [puts small rockwool cube down, goes off set to get more props] I'll get you that cube just to show you it. This is a whole strip of rockwool cubes, [holding strip of rockwool blocks] you break one off. With the stock solution, again adjust the CF level to around 12 then bring the pH down to around 5.5. You can use Formulex again, [picks up bottle] it's tailor-made for rockwool or at this point you can go to Grow A and Grow B nutrient solution, but again when you add the Grow A and Grow B, you have to make it to quite a weak CF level, to approximately 12, which if you was measuring it from the instructions on the back, one third the dilution ratio they tell you on the back. So what you would then do is drop your rockwool cube into the stock solution, [drops rockwool block into solution] allow it to saturate, again you can leave it overnight if you wish, otherwise just let it get to the bottom and then pull it out, [pulls out rockwool block] give it a squeeze to get rid of approximately 10% of the water content from the cube and then without damaging the roots [picks up small rockwool cube and inserts into larger rockwool block] that are already there, ease it into the block and then squeeze it tight. That will then go back into your propagator, ok and you would then water it less, because it is in a much bigger medium, it holds the water more so at that point you'll probably find that you need to water it every 4-5 days instead of every 2-3 days. Ideally, I would then want to propagate with the lid on until you've got a least 3-5 sets of leaves.

Now if you're in a jiffy pellet, and the roots have

grown through the side and bottom, we still recommend going into a bigger rockwool block so although you avoided doing the pH-adjustment and CF-adjustment to get the seedling going, you now have to do it for the bigger rockwool block. [picks up another rockwool block] Once you've done it, the same thing would apply, you would soak it, [drops rockwool block into solution] get it to a level where it is completely saturated, [removes rockwool block] give it a wee squeeze and you put the jiffy pellet in its place. That would then go back in the propagator. Now ideally, you would want to propagate till you've got 3-4 sets of decent leaves. I'm not talking about leader leaves, the original set, I'm talking about decent leaves, so fundamentally what you have is almost a very small plant. At that point you know it's a secure plant even if you shock the life out of it, it shouldn't die. It's at a point now where it's eager to live and strong and it can survive most situations. This is the point if I was a beginner, that I would then transplant it into the system. However, I would wait until roots are growing through the bottom or growing through the sides. Ideally, you do need to see root systems coming through the side or bottom before going into your system. Most systems run on a lot of water, now if the rockwool blocks become absolutely saturated and the rootball isn't through the rockwool block then what will happen is it will starve itself of aeration and prohibit the growth of the plant, so wait until you've got nice roots growing through the side and bottom, 3-4 sets of leaves; at that point it can go into your system.

That is seedling/young plant propagation process finished.

Guide to Hydroponic Systems and Growth Cycles

Presented by Jeffrey Winterborne

The seed has now developed 3-4 sets of leaves; it's got a nice root system in the bigger rockwool block. At that point it's ready to be transferred into your system which basically means you have to get your system ready to receive them.

Now, you have the medium which are clay pebbles. Clay pebbles through transit and the original process of the manufacture of them, they end up quite dirty, quite dusty. Ideally, you want to get rid of all the grit, all the dust, all the debris before putting them into the pods themselves. Now the easy way to do that is simply stab some holes using a big screwdriver into the bottom of them, hose pipe on top, in your garden, for at least half an hour; the longer you do it for, the better. The more dirt, the more grit, the more debris you get off of the pebbles, the better off the pH stability is going to be in your tank. If you haven't got a garden, obviously you've got to use the messy method which means take it into the shower, open the top up, stick some holes in and shower it, or even stick it in your bath and sieve them out. It is fundamental that you do wash the clay, otherwise you end up with a hell of a sediment on the bottom, and every time you disturb the sediment, it could alter the pH which is not what you need. So, wash the clay thoroughly. Once the clay has been washed absolutely thoroughly, it's ready to go into the system.

Before putting it in the system, [picks up one of the pods] make sure that the grommets [points to the pipes in the pods] which are in between the pipes are snugly sat in place. Sometimes in transit, these grommets can come out. Now if you don't have the grommets in place and you put the pebbles, the pebbles can get down the pipes and it can cause all sorts of troubles with the flood and drain cycle. So just make a quick check that the grommets are in the pipes; if all the grommets are in the pipes then fill up the pods with the pebbles. If they are not, find the grommets and put them in their right place. OK, [puts pods down] so we've checked that the grommets are there, the pebbles are washed, the pebbles then go in the top of the tank onto the pods themselves; you really want to fill them up to the level of the overflow pipe, [points to overflow pipe] ok which will be approximately about 12 litres of clay pebbles. Now I'm just going to put these out of the way. [removes sack of pebbles from the table] Now, pods are full with pebbles which have been washed; we're then ready to fill the reservoir up with water.

Now a four pod reservoir holds approximately 45 litres of water. Now these are independently housed reservoirs which means that you can fill it up with a bucket or you can fill it up with a hose. They don't need to be plumbed in, so you'd fill the reservoir up to 45 litres of water; we recommend filling to the overflow pipes. At that point you've got enough space if you need to adjust up and down with water to dilute the solution, you can do so.

OK, we've washed the clay pebbles, we've checked the grommets are in place, we've filled the pods up with clay to the overflow pipes. [picks up pump] Now, we've filled the reservoir up with approximately 45 litres of water, we've then got our pump. The pump connects directly to the down pipe inside the reservoir. It's a submersible pump; a lot of people don't realise that ironically, but that's the way it works, underneath the water. So what we have is the flow pipe attaches directly to the top of the pump and the pump sits in the bottom of the water tank.

Now, [puts pump down] once that's done, you can

then test the flood and drain cycles and it's worth doing this because the bigger the pod system you use, the more time you need to flood the system. On a four pod it should take approximately 1 minute to flood the entire 4 pods, on an eight pod, it should take approximately 3, sometimes it takes 4, on a twelve pod it can take 4, sometimes 5 minutes. On a bigger system where it's using outer runs, you can add another minute on because the outer runs have more pipe which means you need more time to do the flood and drain.

What we always recommend doing is an acid test; basically get the pump in the water, then turn the pump on, time how long it takes to wet your finger, put your finger approximately 1 inch beneath the clay pebble surface and physically time how long it takes before your finger becomes wet. It's a very easy acid test to do but that will give you the precise time you need to do a flood and drain cycle. So, we've got the water in, we've turned the pump on, we're timing how long it takes and we've discovered it takes 1 minute on a 4 pod. But do test it yourself just to confirm it.

Assuming it takes 1 minute to flood the entire system, you then have to program your digital timer. [picks up digital timer] Now this is possibly the hardest bit of the whole equation is programming one of these little beasties. They're very similar to programming VCR recorders from 10 years ago where you have to put the time in, the date in, you then have to program it to come on and off. It does come with full instructions and if you persevere with them you will understand it, however, we are only a phone call away and we don't mind being questioned about it because we're used to it now. They are not too hard once you get your head round it. Quite simply, what you have to do is, if the 4 pod is taking 1 minute to flood, let's say our light cycle is turning on at 9 o' clock in the morning, then what we would also want is our pump cycle to coincide with it, to turn on at 9 o'clock in the morning. So at 9 o'clock the pump turns on, you program that into your timer and then at 1 minute past 9, you program to turn the pump off. What we recommend doing on a system of this design, be it a 4 pod, an 8 pod, a 16 pod or a 24 pod is to flood the system every 3 hours. Every 3 hours until your programs have run out. It should do it basically 24 hours a day.

Now, the system only really needs, or the plants only need water during the lighting cycle, however, we've found that flooding and draining at night does no harm to the plants. However, it does stop the tank from stagnating. It gets oxygen into the water; it keeps the water in motion. The more the water is in motion, the more alive it is. So program your timer to come on every 3 hours for the desired amount of time and the way you find that desired amount of time out is putting your finger in the system, 1 inch beneath the pebbles and timing how long it takes before your finger becomes wet. Just to recap, on a 4 pod it should take 1 minute, on an 8 pod it should take 3 minutes, on a 12 pod it should take 4-5 minutes. And if you've got a 16 or a 24 pod that employs outer runs, you want to add another minute onto the outer run. And that's basically how you would flood and drain the system, but you do have to turn it on and turn it off, otherwise you end up turning your pod system into a deep trough NFT and you lose the benefit of the flood and drain cycles.

Now, once you are happy [puts timer down] that it's flooding and draining to the desired level, every 3 hours, you would then adjust the nutrients in the tank. Now, just like rockwool, you would add the Grow nutrient first before adjusting your pH. At this point, you are not using Formulex – the one part nutrient, you'll be going onto a two part nutrient. [picks up 2 bottles] Now two part basically means you've got Part A and Part B.

Now, they are dual in nature. The reason that you have A and B in separate bottles is because when they are concentrated, they conflict. They don't conflict to the same aggressiveness as pH up and pH down - there's no danger in mixing these 2 together, however, if they do get mixed together in a concentrate, you get nutrient lock which basically means that the plant will not be able to dissolve the desired nutrients it needs to sustain itself. So when measuring A and B out, use separate measurement containers, separate pipettes or separate measuring cups. Never mix these 2 together concentrated otherwise the plant ends up losing out from the nutrients it should get.

Now, what we've got is 45 litres of water in the tank, you would then follow the instructions on the side of the bottle. In this case, it tells us that it requires 4 ml of A and 4 ml of B to each litre of nutrient tank. OK. So you would put 4 ml in per litre of water in the tank and you could measure that out and you will get it somewhere right. Now on these nutrients, on all bottles of nutrients, they tell you the full dosage rate, i.e. if you had a plant that's 2-3 foot tall, strong, in the middle of fruit or flowering, what they're referring to is full strength nutrient on the side of the labels. What you have to do is remember that you are only putting seedlings and cuttings in and they actually desire approximately one third to half the nutrient a fully grown plant would want.

Therefore you would half the dilution ratio it tells you on the side of the bottle. And just to confirm you've got it right, [puts bottle down and picks up meter] you would then employ a TDS meter. Now what I would personally do is pour a splash in, [picks up bottle again] give it a stir, pour exactly an equal amount of B in and give it a stir. Then I would then test with the CF meter until I got to the desired level. Ideally, where you want it is approximately CF of 12, ok, just like the Formulex. You want that level of 12, that will give you just the right amount of nutrient for a young plant and you would keep it on that CF level of 12 until you got the plant approximately 9 inches to 1 foot in height.

So to recap, you would half the dilution ratio, put that into your tank, you would then stir the tank, put the meter in to confirm you've got it to the desired level. If you find that the CF reading is too high, you would then add more water into the tank to dissolve the level in order to reduce the CF value. If you find that the CF level is too low, you would then administer an equal amount of A and an equal amount of B into the tank, give it a stir until you got the level to the right number which in this case for young plants we recommend a CF level of 12. You can go to 14, so 12-14 is approximately where you'll want it. Once you've balanced the nutrient in the tank and you've got it to the desired CF level, oh by the way if you put too much nutrient into the tank you've obviously got to pump a lot of it out to dilute it down. So you would add a hose pipe to your pump, turn the pump on, pump one third of the tank out, then add normal water in order to dilute the level down. So, if it's too low add more, if it's too high, take it out, add water. Simple.

Next, once you've got the level to approximately 12-14 you would then check the pH level. [picks up pH meter] You always do the pH after you do the food because food alters the pH so there's no point getting the levels right with the pH then adding food because you're going to find that that's going to skew-whiff the level. So you always do the pH after you've added the food in the tank. It'll save you a lot of time and a lot of pH up and pH down. So, we'd then pop the pH meter in and again the level that we're looking for now would be a pH of 6. So it's slightly higher than the rockwool because the rootball has now migrated through the rockwool and is now looking to live in the clay pebbles. And ideally, clay pebbles to start with, a pH of 6 is approximately where you want it. So, if you found the pH level is high, you would administer your pH down in very small amounts exactly like we did with the rockwool, always adding a little each time until you get it right.

Now, once you've got it to a pH level of 6, ok, you would then want to flood and drain the system a few times in order to make sure the reservoir is mixed. There's an override button on the timer and all you have to do is switch it on [top button on timer]. You then let the system flood and then you would switch it off, give it a couple of minutes and all the water returns back to the reservoir. It will be a fully mixed tank then. You would then test again with your CF level [picks up CF meter] to make sure that the food level is right and adjust up or down, bearing in mind what the meter says.

After you've done that you would then check your pH [picks up pH meter] and check your pH level, adjusting up or down as you need it. Now because the cuttings or seedlings are very young that are going in the system, if you accidentally put too much pH down in and you end up going to a pH of 5 for example, you're much better off emptying out the tank and starting again than you are using pH up. pH up and down should only be used when a plant is strong enough to cope with them. If you use a lot of pH up and down when your plant is very young – 3 sets of leaves for example – it can do irreparable damage to the rootball. So if you do take the pH down too far, empty out some water, put some water in and that should hopefully raise the level up. Start again basically.

pH up and down are really only to be used for very fine adjustment – never use a lot of it – if you use a lot of it, you end up doing the plants more harm than good. So, once you've got the pH level to 6, the TDS level to 12, the system is flooding and draining, you're ready to put the plants in the system. [take rockwool block from the propagator] What you would do is have a lovely well rooted plant; now you want to administer it, oh let's put that another way. What you want to do is plant it in the system doing minimal damage to the rootball. Now an easy way to do that is to turn the pump on, flood the system so all 4 pods are completely saturated with water. At that point, you can bury the whole rockwool block into the pebbles without doing much damage to the rootball. What you do have to remember is to take the plastic off of the rockwool block. [picks up rockwool block with plastic covering on it] If you don't take the plastic covering off, it will prohibit the rootball from growing out of the side of rockwool; in so doing, restricting the oxygen to the root system. Now that's one way of doing it, is flooding the system and then burying the rockwool in. The other way of doing it is bailing out x amount of clay and then putting it in place and then putting clay on top in order to secure its position. Either way will suffice; I find it a lot easier and quicker just to push it into a flooded system.

Now once it's in place, you'd want a pebble mass just to cover the top of the rockwool itself, i.e. 1-2 levels of pebbles. You do not want the rockwool exposed to the light. Because the rockwool is very water retentive, it ends up holding a lot more moisture and in so doing when the light's reacting with the moisture, you can get algae breakouts which isn't too sightly; it's not that detrimental to the plant but it can encourage all sorts of unwanted diseases and bacteria and that sort of thing. So if you bury it beneath the clay pebbles, the pebbles act as a barrier to the light reacting with the moisture in the rockwool.

Now you would then at that point, look to lower the light possibly by 1 foot. Originally you had the light 3-4 foot away from the canopy of the propagator; at this point now you would want to be 3-2 foot away from the canopy of the plant, so you would raise it or lower it slightly, but you wouldn't want to drop 1 foot instantly ok. If you dropped it 1 foot overnight, the plants could go into shock, so you gradually do it over the course of a week, a couple of inches a day. There's a very easy acid test to find out exactly how close you can get the light to the plant and that's simply the back of your hand. If you raise the back of your hand up to the lamp itself, as soon as you feel any radiant heat hitting the back of your hand, that is approximately how close you can get the light to the canopy without killing it. However, on seedlings and cuttings and young plants, you have to overcompensate that. To give you an example, if you had a plant that is approximately, I don't know, 2 foot tall, and strong and bushy and really having the lust of life, you would then lift your hand up to the back of the light until you felt the radiant heat and then you would get that as close as you can get your hand to it without burning your hand, without feeling that heat.

When they're this big [about 8-12 inches high] you still want the light a good 3 foot away from the canopy of the plant. The plant will then grow up, as the plant is growing up you want to be bringing the light down, so the plant's growing up, the light's coming down till you get to the point where you can't get the light any closer to the plant without burning it, then the plant goes up and so does your light. Now at this point what we recommend doing is raising the plant to approximately a foot, a foot and a half tall on your Grow A and B nutrient, [picks up nutrient bottles] ok, you're on a photoperiod of approximately 18-20 hours a day and you're using the Grow continuously through that period, always keeping the CF level to around 12-14, always keeping the pH level to approximately 6. You can let the pH raise to 6.5 and then knock it back to 6, then let it raise to 6.5 and then knock it back, but always maintain it at around 6.

Now, [picks up bloom bottles] once they're a foot, a foot and a half tall, you can then basically change your nutrient solution to Bloom, or it's also known as Floras; it's also known as Flower which is a different stock solution, but you would also then change your lighting period down to 12 hours on, 12 hours off. Twelve hours on, 12 hours off in most species of plant will induce the fruit or flowering phase. What the plant thinks is happening is the summer is coming to an end, the nights are drawing in and they are forced to reproduce, so what you are basically doing is tricking the plant into thinking that the summer and the winter is coming on strong, i.e. the summer is ending, winter's coming we better reproduce our fruit or flowers and a 12 hour photoperiod in most plants will induce it to fruit or flower.

Now, when you've gone onto the 12 hour light period, you will then want to go onto, instead of Grow nutrient, you'd go onto Bloom nutrient. [picks up bottles again] Bloom nutrient. You have Bloom A and Bloom B. Dual pack nutrient. Again, you would administer the exact amount of A into the reservoir and the exact amount of B in the reservoir. Now you would only use Bloom A and Bloom B when you've got to a plant at least a foot and a half tall and you're on a 12 hour photo cycle, or if the plants are showing signs of fruit or flowering.

Now, at this point the plants need more nourishment, ok, we've been under feeding them on purpose in order to make them hungry, so at this point they can suck up more of what they need. What we would recommend now is to raise the CF level in your tank [picks up CF meter] to approximately 18; so you're going from 12-14 to approximately 18 –20. Ok. So what you would then do is administer more A and more B into your tank. You're almost on full strength as per the instructions on the side of the bottle, but not yet. Almost. Probably about 80-90% strength. In so doing, you would then still maintain the pH level to 6–6.5. Ok.

Now, it's good practice every now and then, approximately 3-4 weeks, to do a nutrient tank flush. To put that in layman's terms, to empty out the tank, put fresh water in and start again. It's not a necessity, however, it is highly advisable. Easy way to empty out the tank is just to attach the hose pipe to the pump, turn the pump on and empty it out. What a lot of hydroponicists would also recommend is running nothing but pH-adjusted water in your system for a day. In so doing, it flushes any dead salts, any unused salts out of the media back into the reservoir, but you do have to adjust the water, pH-adjust it to approximately 6.

So, let's say on this bloom cycle 3-4 weeks in, you said let's get rid of the tank, empty it out and start again. (that make sense?) Right, now it's good practice to do that at least once a month, however, we've known situations where; we've even done it ourselves, where we've not flushed the tank at all, not for the whole crop, not even for 2 crops. You can get away with not doing it, however, puritan hydroponicists would say, aagh, do it every 2 weeks, do it every 3 weeks and it's not a necessity, but it is, well it can help the plants but if you are lazy, you don't have to.

Now, we've put Bloom A and B into the tank to get to a CF/TDS level of 18-20, we've then adjusted the tank to 6 pH level and we're still flooding and draining. At this point, we've the got the light as close as we can to the plants without burning them. Now what you will find is that some plants are going to grow quicker than other plants. It's the law of nature. You always get some that are stronger than others. The ones that are taller, we recommend highly, pinching them out. Either chopping the tops off or taking the centre 2 growth, the centre tip out of the plant. In so doing, it encourages the side growth of the plant instead of the centre growth. That will allow the smaller plants to catch up in height, and again keep on doing it.

The name of the game in hydroponics is to get that light as close as you can to ALL the plants.

If you get a few that bolt, that are growing taller than the others, you have to chop them back. You've got to be cruel to be kind. Because if they bolt and you end up lifting the light up, because they're growing so tall, the others are then going to elongate and stretch because the light isn't close enough to them, and in so doing, the tall plants will benefit and the little plants won't be worth growing at all.

OK, now the plants are going to continuously grow and the light is going to continuously go up with the plants' growth. Ideally, you want to keep the crop even, the light down as close as you can to all the plants and let the plants grow and the light evolve as the plants grow. Now approximately 5 weeks into the Bloom cycle, into the fruiting cycle, i.e. the 12 hour lighting scheme, you would want to administer some solution called PK13/14; it's a bloom stimulant, fruiting boost stimulant.

Now plants only require this PK13/14 during the 5th week of the 12 hour lighting cycle, so when you go into your blooming stage when you start using your bloom food and you first initiate your 12 hour light cycle, you must write down the date because 5 weeks into that cycle, you want to add to your nutrient tank the PK13/14. Some species of plants take longer to fruit or flower. When you administer the PK13/14, it's approximately 3 weeks before harvest of your plants. You'd follow the instructions on the back of your bottle and go in

full strength, but never exceeding a CF level [picks up CF meter] of 22-24. So you'd add your stock solution, your Bloom, you'd then add your PK13/14, get it to a level of around 22-24. If it's below that level, you'd add more Bloom A and B. Ok. If you do end up administering too much nutrient into the tank, bleed some of it off and start again using the PK13/14 first as your primary nutrient. Again, once that's in the tank, reduce the pH level down to approximately 6. You would have used the PK13/14 or bloom stimulant for approximately 1 week, ok, so once you've added it to the tank, you use it for the 1 week and 1 week only. You wouldn't administer it again.

You will find at this point, 3 weeks before harvest, the plants are going to be sucking up a whole tank at least a tank a week. So, you're going to be filling up that tank possibly on a daily, or a bi-daily basis. And once you've filled it up over the course of the week, then go back onto your [picks up bottles] Bloom A and your Bloom B and not using your PK13/14 anymore. At that point, when you're 1 week away from harvesting, 1 week away from cropping your plants, what we highly recommend doing is emptying out the reservoir, filling the reservoir up with nothing but clean water, pH-adjust it to a pH of 6 and then feed the plants nothing but clean water for 1 week. This is what they call flushing.

It benefits multi levels. One, it benefits the system because it gets rid of any nutrient build up in the pebbles themselves. Two, it benefits the plant because it actually encourages the plant to take all the nutrient out of it, itself. If you're growing fruit or veg, or any plant that is consumed, if you don't flush it at the end in a hydroponic system, you end up [picks up bottles] with an excessive amount of nutrient build up in the plant and that can make the plant taste quite bitter. So it's fundamental to flush the system and the plants for 1 week before harvest and that point you're ready to crop the plants and start again. Job done.

Once you've harvested your plants, you end up with a lot of root system in your pebbles. Now what we recommend doing is flooding the system, pulling the majority of the rootball out. You can empty out the pods and disinfect it but there's an easier way round that. You take the majority of the rootball out, you would then add a nutrient to the tank called Cannazine. Cannazine has active enzymes that dissolve dead root systems. They turn those root systems into food for the next crop. So if you don't want to get all messy and sweaty, take the majority of the rootball out, and then put a good strong dose of Cannazine into the tank and flood and drain it every 3 hours like you used to. You can then plant on top of that existing root system as well as not having to worry about changing the reservoir. We do, however. recommend before you use the Cannazine is to use a very strong H_20_2 hydrogen peroxide bleach to disinfect the system in between crops. Disinfecting it will discourage any diseases of any description. So, firstly, remove the rootball or the majority of it, you then add H_20_2 to the tank, you flood and drain that, you then bleed that H_20_2 off, add new water using Cannazine and then at that point you can plant on top of your existing system. And that's how you would start again.

A www.1-hydroponics.co.uk Production

Hydroponics
Indoor Horticulture

Appendix 2

Useful Information

What follows is various and very useful information for reference purposes when constructing and maintaining your grow room.

Lineal Measurements

Common Use Imperial
- 12 inches = 1 foot
- 3 feet = 1 yard
- 1760 yards = 1 mile

Other Imperial Values
- 12 inches (ins or ") = 1 foot (ft or ')
- 3 feet = 1 yard (yd)
- 22 yards = 1 chain (ch)
- 10 chains = 1 furlong (fgs)
- 8 furlongs = 1 mile (mi)

Common Metric Measure
- 10 millimetres (mm) = 1 centimetre (cm)
- 1 metre (mtr) = 1000 millimetres
- 1 metre = 100 centimetres
- 1000 metres = 1 kilometre (km)

Conversions
- 1 inch = 25.4 millimetres
- 1 inch – 2.54 centimetres

- 1 foot = 304.8 millimetres
- 1 foot = 30.48 centimetres
- 1 foot = 0.3048 metres

- 1 yard = 9144 millimetres
- 1 yard = 914.4 centimetres
- 1 yard = 0.9144 metres

- 1 mile = 1609.34 metres
- 1 mile = 1.60934 kilometres

- 1 millimetre = 0.03937 inches
- 1 centimetre = 0.3937 inches
- 1 metre = 39.37 inches
- 1 metre = 3.28084 feet
- 1 metre = 1.0936 yards
- 1 kilometre = 39,370 inches
- 1 kilometre = 3280.84 feet
- 1 kilometre = 1093.613 yards
- 1 kilometre = 0.621317 miles

Area Measurements

Common Use Imperial
- 144 sq ins (sq") = 1 sq ft (sq')
- 9 sq' = 1 sq yd
- 4840 sq yds = 1 acre
- 640 acres = 1 sq mile

Common Metric Measure
- 10 mm^2 = 1 cm^2
- 10,000 cm^2 = 1 m^2
- 1,000,000 m^2 = 1 hectare (h)

Conversions
- 1 sq in = 6.45160 cm^2
- 1 sq ft = 0.09290 m^2
- 1 sq yd = 0.83613 m^2
- 1 acre = 0.40469 h
- 1 sq mile = 2.58999 km^2
- 1 cm^2 = 0.155 sq"
- 1 m^2 = 10.76391 sq'
- 1 m^2 = 1.19599 sq yds
- 1 h = 2.47105 acres
- 1 km^2 = 0.38610 sq miles

Volume Measurements

Common Use Imperial
- 1728 cu ins = 1 cu ft
- 27 cu ft = 1 cu yd

Common Metric Measure
- 1000 mm^3 = 1 cm^3
- 1,000,000 cm^3 = 1 m^3

Conversions
- 1 cu in = 16.38706 cm^3
- 1 cu ft = 0.02832 m^3
- 1 cu yd = 0.76455 m^3
- 1 cm^3 = 0.061024 cu ins
- 1 m^3 = 35.31467 cu ft
- 1 m^3 = 1.30795 cu yds

Capacity Measurements

Common Use Imperial
- 20 ounces (oz) = 1 pint (pt)
- 2 pts = 1 quart (qt)
- 8 pts = 1 gallon (gall)
- 4 qts = 1 gall

Common Metric Measures
- 1000 millilitres (mls) = 1 litre (ltr)

Conversions
- 1 pint = 0.56826 ltrs
- 1 qt = 1.13652 ltrs
- 1 gall = 4.54609 ltrs

NB: 1 US gall (USA) = 3.7854 ltrs

- 1 ltr = 1.75976 pts
- 1 ltr = 0.87988 qt
- 1 ltr = 0.2641728 US gall

Mass Measurements

Common use Imperial
- 16 ounces (oz) = 1 pound (lb)
- 14 lbs = 1 stone (st)
- 28 lbs = 1 quarter (qtr)
- 4 qtrs = 1 hundredweight (cwt)
- 112 lbs = 1 cwt
- 20 cwt = 1 ton (t)

Common Metric Measure
- 1000 grams(gm) = 1 kilogram (kg)
- 1000 kg = 1 metric ton (tonne)

Conversions
- 1 oz = 28.349532 gm
- 1 lb = 0.453592 kg
- 1 cwt = 50.80234 gm
- 1 T = 1016.0469 kg
- 1 T = 1.0160469 tonnes

Power Measurements

Common Use Imperial
= the horsepower (hp)

Common Metric Measure
= the kilowatt (Kw)

Conversions
- 1 hp = .7457 Kw
- 1 Kw = 1.341022 hp

Other measurements and inter-relationships of energy, work and power are

- 1 foot pound = 0.1382 kilogrammetre = 0.00129 Btu = 0.000376 watt-Hr = 1.36 joules = 0.000326 calories

- 1 kilogrammetre = 7.233 ft lb

- 1 foot ton = 0.309 tonne metre

- 1 tonne metre = 3.23 ft tons

- 1 hp = 33,000 ft lb per minute = 746 watts = 42.4 Btu per min = 1.014 metric hp

- 1 force de cheval = 0.9863 hp

- 1Kw = 1000 watts = 1.341 hp = 56.9 Btu per minute

- 1 watt = 44.36 ft lb per minute

- Volts x amperes = watts

- 1000 watts = 1 Kw
- 1000 watts for 1 hour = 1 kWh

- 1 British thermal unit = the heat required to raise 1 lb of water 1 degree Fahrenheit

- 1 Btu = 778.26 ft lb = 0.2931 watt Hr = 1043 joules = 0.000393 hp Hr = 107.5 kilogrammetres = 0.252 calories

- 1 Kilowatt hour (kWh) = 3413 Btus

Temperature

To convert degrees centigrade (Celsius, ˚C) to degrees Fahrenheit:

(˚F) multiply x 9, divide by 5 and add 32
example:
100˚C x 9 = 900 ÷ 5 = 180 + 32 = 212˚F

To convert degrees Fahrenheit to degrees Centigrade subtract 32, then multiply x 5 and divide by 9:
example:
176˚F – 32 = 144 x 5 = 720 ÷ 9 = 80˚C

Water Heating Data

Note that this data refers to the heating of pure water. Some small differences will actually occur when heating nutrient, however, for the purposes of sizing heating equipment, these differences can be ignored.

- 1 gallon = 227 cubic inches
- 1 gallon of fresh water = 10 pounds
- 1 cubic foot = 6.23 gallons
- 1 cubic foot fresh water = 62.3 lb
- 1 Imp gallon = 4.54 ltr
- 1 ltr = 0.22 Imp gall
- 1 inch water column = 0.036 lb per sq in = 5.194 lb per sq ft
- one foot head of water = 0.433 lb per sq in = 62.35 lb per sq ft
- one lb per sq in = 27.71 in head of water = 2.04 inches mercury

- 1 atmosphere @ 60˚F = 14.7 lb per sq in (2116.3 lb per sq ft) = 30” mercury (33.96 ft water)

- 1 in mercury = 13.62 in water = 0.49 lb per sq in

Water expands approximately 4% when heated from cold to boiling.

To calculate the holding capacity in gallons of a rectangular container:

Multiply length x breadth x height (in inches) and divide by 227 or multiply in feet and divide by 6.23

To calculate the holding capacity in gallons of a cylindrical container:

Multiply height x diameter x diameter (in inches) and divide by 350.

(use the earlier data to convert from Imp galls to US galls or litres)

To calculate:

1) kWh = (galls x 10 x temp rise ˚F) divided by (3423 x efficiency)
2) kW = (galls x 10 x temp rise ˚F) divided (3413 x efficiency x time in hours)
3) galls/hour = (kW x 3413 x efficiency) divided by (temp rise in ˚F x kW)
4) time in hours = (galls x 10 x temp rise in ˚F) divided by (3413 x efficiency x kW)

Working example:
22 gallons, 90˚F temp rise, 0.8 efficiency
kWh = (22 x 10 x 90) divided by (3413 x 0.8) = (19800) divided by (2730.4) = 7.26

For similar calculations using metric values:

1 kilogram calorie = heat required to raise the temperature of 1 litre of water 1 degree centigrade

- 1 kilowatt hour = 860 kilogram calories
- 1 Btu = 0.252 calories
- 1 calorie = 3.9683 Btus

To calculate:

1) kWh = (litres x temp rise ˚C) divided by (860 x efficiency)
2) kW = (litres x temp rise ˚C) divided by (860 x efficiency x time in hours)
3) ltr/hour (kW x 860 x efficiency) divided by (temp rise ˚C)
4) Time in hours = (ltr x temp rise ˚C) divided by (860 x efficiency x kW)

Working example:
100 litres water, 50˚C temp rise, 0.8 efficiency

kWh = (100 x 50) divided by (860 x 0.8) = (5000)/(688) = 7.267

From a heating perspective, the efficiency of a nutrient reservoir will depend upon its ability to conserve the heat which is applied. Good insulation around the reservoir can greatly improve this efficiency. As an approximate guide, efficiencies of 0.9 to 0.95 could be expected with proper insulation.

Calculations of heating requirements are a direct relationship to the loss of heat from the grow room environment. Heat loss will be via the roof and walls.

To calculate the heat input required to maintain the desired temperature, it is necessary to calculate the surface area of the grow room. This is achieved by calculating each rectangular panel, say 2 side walls + 2 end walls + 4* triangular sections of end walls + 2 roof sections = total area of hip roofed structure.

Rectangles are calculated by multiplying the length by the breadth = area.

It will possibly be easier for the less mathematically minded to calculate the end triangular sections as 2 right angled triangles. ☺

Triangles, call the 3 sides A, B and C

Area of right angle triangle = $A \div 2 \sqrt{C^2 - A^2} = AB \div 2 = B^2 \tan A \div 2$

More Useful Information

- 1 ton of displacement (i.e. a boat) = 35 cubic feet.
- 1 Imperial gallon of distilled water weighs 10 lbs.
- A standard wine glass holds approximately 2 fluid ozs.
- A teacup holds approximately 3 fluid ozs.

Decimal and Millimetre equivalents

- 1/64 in = 0.1625 in = 0.3969 mm
- 1/32 in = 0.03125 in = 0.7937 mm
- 3/64 in = 0.046875 in = 1.1906 mm
- 1/16 in = 0.0625 in = 1.5875 mm
- 5/64 in = 0.078125 in = 1.9844 mm
- 3/32 in = 0.09375 in = 2.3812 mm
- 7/64 in = 0.109375 in = 2.7781 mm
- 1/8 in = 0.125 in = 3.1750 mm
- 9/64 in = 0.140625 in = 3.5719 mm
- 5/32 in = 00.15625 in = 3.9687 mm
- 11/64 in = 0.171875 in = 4.3656 mm
- 3/16 in = 0.1875 in = 4.7625 mm
- 13/64 in = 0.203125 in = 5.1594 mm
- 7/32 in = 0.21875 in = 5.5562 mm
- 15/64 in = 0.234375 in = 5.9531 mm
- 1/4 in = 0.25 in = 6.3500 mm
- 5/16 in = 0.3125 in = 7.9375 mm
- 3/8 in = 0.375 in = 9.5250 mm
- 1/2 in = 0.5 in = 12.7000 mm
- 5/8 in = 0.625 in = 15.8750 mm
- 3/4 in = 0.75 in = 19.0500 mm
- 7/8 in = 0.875 in = 22.22250 mm

- 1 in = 25.4 mm
- 2 in = 50.8 mm
- 3 in = 76.2 mm
- 4 in = 101.6 mm
- 5 in = 127.0 mm
- 6 in = 152.4 mm
- 7 in = 177.8 mm
- 8 in = 203.2 mm
- 9 in = 228.6 mm
- 10 in = 254.0 mm
- 11 in = 279.4 mm
- 12 in = 304.8 mm

- 1 mm = 0.03937 in
- 2 mm = 0.07874 in
- 3 mm = 0.11811 in
- 4 mm = 0.15748 in
- 5 mm = 0.19685 in
- 6 mm = 0.2362 in
- 7 mm = 0.27559 in
- 8 mm = 0.31496 in
- 9 mm = 0.3433 in
- 10 mm = 0.39370 in
- 20 mm = 0.78740 in
- 30 mm = 1.18110 in
- 40 mm = 1.57480 in
- 50 mm = 1.96850 in
- 60 mm = 2.36220 in
- 70 mm = 2.75590 in
- 80 mm = 3.14960 in
- 90 mm = 3.54330 in
- 100 mm = 3.93700 in
- 200 mm = 7.87401 in
- 300 mm = 11.81102 in
- 400 mm = 15.74803 in
- 500 mm = 19.68503 in
- 600 mm = 23.62204 in
- 700 mm = 27.55905 in
- 800 mm = 31.49606 in
- 900 mm = 35.43307 in
- 1000 mm = 39.37007 in

The conductivity of a nutrient solution can be compared to a known weight of chemical being added to a known volume of water. The result is dependent upon a) the quality of the water, and b) the salt being added i.e. the conductivity will vary according to the particular nutrients being used.

CF units to grams per litre (g/Ltr)

CF Units	g/ltr	CF Units	g/ltr
1.1	0.1	19.8	1.8
2.2	0.2	20.9	1.9
3.3	0.3	22.0	2.0
4.4	0.4	27.5	2.5
5.5	0.5	33.0	3.0
6.6	0.6	44.0	4.0
7.7	0.7	49.5	4.5
8.8	0.8	55.0	5.0
9.9	0.9	60.5	5.5
11.0	1.0	66.0	6.0
21.1	1.1	71.5	6.5
13.2	1.2	77.0	7.0
14.3	1.3	82.5	7.5
15.4	1.4	88.0	8.0
16.5	1.5	93.5	8.5
17.6	1.6	99.0	9.0
18.7	1.7		

The variation of nutrient salts to be used in preparing a hydroponic nutrient formula depends on several factors.

1. Relative proportions of the ions within a compound must be compared with that required in the nutrient formulation
2. Solubility of the nutrient salts i.e. calcium could be provided by either calcium nitrate or calcium sulphate. Calcium sulphate is less expensive but it is only partially soluble and therefore does not have a place in hydroponic formulae
3. Cost of material is also a factor to consider. Commercial fertiliser grades of some ingredients are suitable, however, the purity of the materials can sometimes be lacking, so these attributes must be known before use since an impurity can lead to a toxicity of a trace element if present in excessive quantities. Substandard grades can have insoluble inert carriers such as clay or silt particles which will form a sludge, and in the extreme, can cause nutrient lock.

Chemical Formula	Chemical Name	Elements Supplied	Solubility Ratio
KNO3	Potassium Nitrate	K+, NO3-	1:4
Ca(NO3)2	Calcium Nitrate	Ca++, 2(NO3)-	1:1
KH2PO4	Mono Potassium Phosphate	K+, H2PO4-	1:3
KC1	Potassium Chloride	K+, CI-	1:3
K2SO4	Potassium Sulphate	2K+, SO4 =	1:15
Mg SO4.7H2O	Magnesium Sulphate	Mg++, SO4 =	1:2
CaCI2.6H2O	Calcium Chloride	Ca++, 2CI -	1:1
CaSO4.2H2O	Calcium Sulphate	Ca++, SO4 =	1:500
H2PO4	Phosphoric Acid	PO4=-	Acid Soln
FeEDTA	Iron Chelate	Fe++	High
H3BO3	Boric Acid	B+++	1:20
MnSO4.4H2O	Manganese Sulphate	Mn++, SO4 =	1:2
CuSO4.5H2O	Copper Sulphate	Cu++, SO4 =	1:5
ZnSO4.7H2O	Zinc Sulphate	Zn++, SO4 =	1:3
(NH4)6Mo7O24	Ammonium Molybdate	NH4+, Mo+6	1:2.3

Here are the common ways in which the units of electrical conductivity are expressed:

1	Electrical conductivity	'EC'
2	Conductivity factor	'CF'
3	Microsiemens/cm	'mS cm-1' or 'mS'
4	millimhos	'mmho'
5	micromhos	'mmho'
6	millisiemens	'mS'

These are absolute values based upon scientific scales, and they are all exactly commensurate to one another.

Each is a measurement of 'electrical conductance', which is conversely proportional to the measurement of 'electrical resistance'.

Their interrelationship is:
1 mS =
1 mmho =
1000 mS =
1000 Mmhos =
10 CF =
1.0 EC =

Electrical conductivity readings change with the temperature of the nutrient by approximately 2% per degree centigrade. Therefore, it is important to make sure that the conductivity meter is temperature compensated to take account of this fact.

Saturation Level of Water with Atmospheric Oxygen in mg of O_2/L

t°:	0.0:	0.1:	0.2:	0.3:	0.4:	0.5:	0.6:	0.7:	0.8:	0.9:
10	11.35	11.32	11.30	11.27	11.25	11.22	11.20	11.17	11.15	11.12
11	11.10	11.08	11.05	11.03	11.00	10.98	10.96	10.93	10.91	10.88
12	10.86	10.84	10.81	10.79	10.76	10.74	10.72	10.69	10.67	10.64
13	10.62	10.60	10.57	10.55	10.53	10.50	10.48	10.46	10.44	10.41
14	10.39	10.37	10.35	10.33	10.31	10.28	10.26	10.24	10.22	10.20
15	10.18	10.16	10.14	10.12	10.10	10.07	10.05	10.03	10.01	9.99
16	9.97	9.95	9.93	9.91	9.89	9.86	9.84	9.82	9.80	9.78
17	9.76	9.74	9.72	9.70	9.68	9.66	9.64	9.62	9.60	9.58
18	9.56	9.54	9.52	9.50	9.48	9.46	9.45	9.43	9.41	9.39
19	9.37	9.35	9.33	9.32	9.30	9.28	9.26	9.24	9.23	9.21
20	9.19	9.17	9.16	9.14	9.12	9.10	9.09	9.07	9.05	9.04
21	9.02	9.00	8.99	8.97	8.95	8.93	8.92	8.90	8.88	8.87
22	8.85	8.83	8.82	8.80	8.78	8.76	8.75	8.73	8.71	8.70
23	8.68	8.66	8.65	8.63	8.62	8.60	8.58	8.57	8.55	8.54
24	8.52	8.50	8.49	8.47	8.46	8.44	8.43	8.41	8.40	8.38
25	8.37	8.35	8.34	8.32	8.31	8.29	8.28	8.27	8.25	8.23
26	8.22	8.21	8.19	8.18	8.16	8.15	8.14	8.12	8.11	8.09
27	8.08	8.07	8.05	8.04	8.02	8.01	8.00	7.98	7.97	7.95
28	7.94	7.93	7.91	7.90	7.88	7.87	7.86	7.84	7.83	7.81
29	7.80	7.79	7.77	7.76	7.75	7.73	7.72	7.71	7.70	7.00

Some US publications state the strength of the nutrient solution in ppm (parts per million). While this does provide some guidelines to the strength of the nutrient solution, it cannot be relied upon as an accurate or consistent method for the reasons listed below:

Different elements will be required in greater or lesser amounts in order to get comparable readings and it is this electrical strength that the plant responds to! Using the 'known industry standard' KC1 potassium chloride solution, the following conductivity which is what the plant is actually interested in, the ppm is charted. The table shows the amount ppm of KC1 required at the specific temperature to result in a conductivity value of 1.0EC/10CF

Conductivity: ppm of KCI solution required @ °C to = 1.0EC (10CF):

0.1 EC	15°	16°	17°	18°	19°	20°	21°	22°	23°	24°	25°	27°	29°
(1 CF)	787	767	750	732	715	700	685	671	657	643	631	607	584

Nutrient Element: @ 1 x Millimole per litre equals 'Conductivity':

Nitrogen as	NO3	14 ppm	0.2 CF
Nitrogen as	NH4	14 ppm	0.2 CF
Phosphorus	P	31 ppm	0.44 CF
Potassium	K	39 ppm	0.56 CF
Calcium	Ca	40 ppm	0.57 CF
Magnesium	Mg	24 ppm	0.34 CF
Sodium	Na	23 ppm	0.33 CF
Chlorides	CI	35.5 ppm	0.51 CF
Sulphates	SO4	32 ppm	0.46 CF

Nutrient element: @ 1 x Micromole per litre:

Iron	Fe	0.057 ppm
Copper	Cu	0.064 ppm
Zinc	Zn	0.05 ppm
Manganese	Mn	0.055 ppm
Boron	B	0.1 ppm
Molybdenum	Mo	0.096 ppm

Please note that at 20°C, a content of 700 ppm is required to result in an EC of 1.0 (10CF). This is the 'standard' relationship that many laboratories use when referring ppm to conductivity. A direct relationship could be made at any temperature, for example at 25°C, the same conductivity value would now require only 630 ppm. Let us now assume we are preparing a solution where a US book calls for 750 ppm. We would first need to know at what temperature the author was referring to when quoting the 750 ppm. For example, if it was expressed at 15°C (59°F), the conductivity would be 0.95 EC (9.5CF), whereas at 20°C, it would be 1.08 EC (10.8CF), and at 30°C, it would be 1.3 EC (13 CF).

As if this wasn't enough, you also have to be aware that every time you change the nutrient element involved, you also change the ppm required to give the same conductivity. This makes it almost an impossibility for the hydroponic grower to ensure the optimum strength of nutrient solution is being applied to plants when basing calculations upon an aggregated ppm scale.

In science, to be totally correct, we would refer to the 'molecular weight' (based upon the 'atomic' weight) of an element in solution, expressed as 'Molecular' or 'Molar' Volume:

1 x Mole = 1000 millimoles and 1 x Millimole = 1000 Micromoles

Expressions such as ppm or mg/1 or mg/ml can be converted to 'mole' equivalents. (refer to table)

Nominal CF Values for Hydroponic Crops

African violet	10 to 12	Lavender	10 to 14
Asparagus	14 to 18	Leek	16 to 20
Avocado Pear	18 to 26	Lettuce (Fancy)	03 to 08
Balm	10 to 14	Lettuce (Iceberg)	06 to 14
Banana	18 to 22	Melons	10 to 22
Basil	10 to 14	Mint	10 to 14
Beans	18 to 25	Mustard/Cress	12 to 24
Beetroot	14 to 22	Onion	18 to 22
Blueberry	18 to 20	Parsley	08 to 18
Borage	10 to 14	Passionfruit	16 to 24
Broccoli	14 to 24	Pea	14 to 18
Brussels sprout	18 to 24	Pumpkin	14 to 24
Cabbage	14 to 24	Radish	12 to 22
Capiscum	20 to 27	Rhubarb	16 to 20
Carnation	12 to 20	Roses	18 to 26
Carrot	14 to 22	Sage	10 to 16
Cauliflower	14 to 24	Spinach	18 to 35
Celery	15 to 24	Silverbeet	18 to 24
Chives	12 to 22	Squash	18 to 24
Cucumber	16 to 24	Strawberry	18 to 25
Dwarf roses	16 to 26	Thyme	12 to 16
Eggplant	18 to 22	Tomato	22 to 28
Endive	08 to 15	Turnip, Parsnip	18 to 24
Fennel	10 to 14	Watercress	04 to 18
Kohlrabi	18 to 22		

The plant does not distinguish the individual nutrient elements strength as ppm, rather it distinguishes their electrical strength (conductivity). Therefore, given advice to grow a particular crop at a specific ppm/temp, the best thing to do is to convert that figure to a conductivity value and continue to monitor the solution and add additional nutrients upon the basis of temperature compensated conductivity readings.

Electricity Costs for HID Lights Based on 1 Unit of Electricity @ Pence Per Unit

Wattage	Amps	Inrush Current	Cost		Per day	Per week
1000W	4.3A	7.81A	6.5 pence hour	18 hr	117 p	819 p
				12 hr	78 p	546 p
600W	2.60A	4.68A	3.9 pence hour	18 hr	70.2 p	491 p
				12 hr	46.8 p	327 p
400W	1.73A	3.11A	2.6 pence hour	18 hr	468 p	327 p
				12 hr	31.2 p	218 p
250W	1.08A	1.94A	1.62 pence hour	18 hr	29.1 p	203 p
				12 hr	19.4 p	135 p
M/Blend 500W	2.17A		3.25 pence hour	18 hr	58.5 p	409 p
				12 hr	39 p	273 p
M/Blend 250W	1.08A		1.62 pence hour	18 hr	29.1 p	203 p
				12 hr	19.44 p	135 p
High fluoro 110W	0.47A	0.85A	0.6 pence hour	18 hr	10.8 p	75.6 p
				12 hr	7.2 p	50.4 p

Conversion Charts and Tables

Carbon Dioxide Facts and Figures

Molecular weight = 44 grams/mole
Sublimes (solid to gas) at 78.5°C at 1 atmosphere – air density = 1.2928 grams/litre (i.e. at equal temperatures and pressures carbon dioxide is heavier than air and CO_2 will fall to the bottom of an air/CO_2 mixture).

To calculate a new volume if only previous volume, temperature and pressure are known use the formula:

$$V2 = \frac{T2}{P2} \times \frac{P1}{T1} \times V1$$

Where V2 = new volume in litres
V1 = old volume in litres
P1 = old pressure in atmospheres
P2 = new pressure in atmospheres
T1 = old temperature in degrees Kelvin
T2 = new temperature in degrees Kelvin

If the weight of gas is known use:

Weight CO_2 x 0.08205 x temp. (degrees K)
Volume (litres) =
44 x pressure in atmospheres
(weight of CO_2 measured in grams)

example:

Weight CO_2 = 5 kgs = 5000 grams
Pressure = 14.7 PSI = 1 atmosphere
Temperature = 25°C = 25 + 273 = 298 degrees K

The volume of gas will be
5000 x 0.08205 x 298

44 x 1 = 2.778 litres
= 27.78 cubic metres

Physical properties of Propane:

Formula	C3h8
Boiling point	-44°F
Specific gravity of gas (air = 1)	1.50
Specific Gravity of liquid @ 60°F (water = 1)	0.504
Latent heat vaporation total/bal	773.0
Pounds per gallon of liquid @ 60°F	4.23
Gallons per pound of liquid @ 60°F	0.236
BTU per cubic foot of gas @ 60°F	2488
BTU per lb of gas	21548
BTU per gallon of gas @ 60°F	90502
Lower limit of flammability (% of gas)	2.15
Upper limit of flammability (% of gas)	9.60
Cubic feet of gas per gallon (% of gas)	36.38
Octane number	100+

Combustion Data:

Cubic feet of air to burn 1 gal of propane	873.6
Cubic feet of CO_2 per gal of propane burned	109.2
Cubic feet of nitrogen per gal of propane burned	688
Pounds of CO_2 per gal of propane burned	12.7
Pounds of nitrogen per gal of propane burned	51.2
Pounds of water vapour per gal of propane burned	6.8
1 pound of propane produces in KWH	6.3
BTU's per KW hour	3412
BTU input boiler horsepower	45,000
1 MCF of natural gas (11 gal propane)	1000,000BTU
Therm (1.1 gal propane)	100,000BTU

BTU Content	Per Gall	Per LB	LP Gas Properties	Propane
Propane	91,300	21,600	BTU per cubic ft	2,516
Fuel oil	135,435	16,200	Pounds per gallon	4.24
Liquefied Nat Gas	86,000	23,200	Cubic ft per gallon	36.39
Soft coal		14,000	Cubic ft per pound	8.58
			Specific gravity of vapour	1.52

1 Therm	100,000 BTU	Specific gravity of liquid	0.509
1 cu ft Nat Gas	1,000 BTU	Vapour pressure (psig) 00F	23.5
1 lb steam	970 BTU	Vapour pressure (psig) 700F	109
1 kilowatt	3.413 BTU	Vapour pressure (psig) 1000F	172

Calculations for Metric Users

1 cubic metre = 1m x 1m x 1m = 1,000 litres

Fans are rated at litres per minute or litres per second

Measure room length, width and height in metres
eg 3m x 3m x 2.4m = 21.6 cubic metres

Buy a fan that will clear the grow room volume of air in 1-5 minutes. Run the fan for twice the time to clear the grow room of air.

Work out the amount of CO_2 gas to add
e.g. want 1,500 ppm – ambient is 350 ppm – need to add 1500 – 350 = 1150 ppm

Grow room size in cubic metres

x ppm of CO_2 add = litres / 1000

Most grow rooms probably have a 20% leakage factor which has to be added to the CO_2 gas required.

For our grow room of 21.6 cubic metres and 1500 ppm of CO_2, we need to add

21.4 x 1150 = 24.61 litres x 1.2 = 29.53 litres / 1000

set the flow meter to 6 litres per minute and run the gas for 5 minutes

If you are concerned with the possible loss of CO_2 via extraction in your grow room, then leave the gas-enriched air for 20 minutes and then exhaust the air from the grow room and start the cycle again.

A short range timer programmable down to 1 minute and up to 54 cycles per day, with two of these, one can set up a reliable CO_2 injection system – one timer working the inlet and exhaust fans, and the other timer working the solenoid. This can make a simple mechanism for CO_2-enrichment.

Metric Conversion Chart

When You Know	Multiply by	To Find
Length		
Millimetres	0.04	Inches
Centimetres	0.39	Inches
Metres	3.28	Feet
Kilometres	0.62	Miles
Inches	25.40	Millimetres
Inches	2.54	Centimetres
Feet	30.48	Centimetres
Miles	1.16	Kilometres
Area		
Square centimetres	0.16	Square inches
Square metres	1.20	Square yards
Square kilometres	0.39	Square miles
Hectares	2.47	Acres
Square inches	6.45	Square centimetres
Square feet	0.09	Square metres
Square yards	0.84	Square metres
Square miles	2.60	Square kilometres
Acres	0.40	Hectares
Volume		
Millilitres	0.20	Teaspoons
Millilitres	0.60	Tablespoons
Millilitres	0.03	Fluid ounces
Litres	4.23	Cups
Litres	2.12	Pints
Litres	1.06	Quarts
Litres	0.26	Gallons
Cubic metres	35.32	Cubic feet
Cubic metres	1.35	Cubic yards
Teaspoons	4.93	Millilitres
Tablespoons	14.78	Millilitres
Fluid ounces	29.57	Millilitres
Cups	0.24	Litres
Pints	0.47	Litres
Quarts	0.95	Litres
Gallons	3.790	Litres
Mass and Weight		
Grams	0.035	Ounce
Kilograms	2.21	Pounds
Ounces	28.35	Grams
Pounds	0.45	Kilograms

- 1 inch (in) = 25.4 millimetres (mm)
- 1 foot (12 in) = 0.3048 metres (m)
- 1 yard (3ft) = 0.9144 metres
- 1 mile = 1.60937 kilometres
- 1 square inch = 645 square millimetres
- 1 square foot = 0.0929 square metres
- 1 square yard = 0.8361 square metres
- 1 square mile = 2.59 square kilometres

Liquid Measure Conversion

- 1 pint (UK) = 0.56824 litres
- 1 pint dry (US) = 0.55059 litres
- 1 pint liquid (US) = 0.47316 litres
- 1 gallon (UK) (8 pints) = 4.5459 litres
- 1 gallon dry (US) = 4.4047 litres
- 1 gallon liquid (US) = 3.7853 litres
- 1 ounce = 28.3495 grams
- 1 pound (16 ounces) = 0.453592 kilograms
- 1 gram = 15.4325 grains
- 1 kilogram = 2.2046223 pounds
- 1 millimetre = 0.03937014 inches (UK)
- 1 centimetre = 0.3937014 inches (UK)
- 1 metre = 3.280845 feet (UK)
- 1 kilometre = 0.6213722 miles

Light Conversion

- 1 foot candle = 10.76 = lux
- 1 lux = 0.09293 foot candles
- Lux = 1 lumen/square metres

Celsius to Fahrenheit

- Celsius temp = (F – 32) x 5 / 9
- Fahrenheit temp = (C x 9.5) + 32

Celsius	Fahrenheit
32	89.6
30	86
28	82.4
26	78.8
24	75.2
22	71.6
20	68
18	64.4
16	60.8
14	57.2
12	53.6
10	50
8	46.4
6	42.8
4	39.2
2	35.6
0	32

Humidity

Relative humidity is the ratio factor between the amount of moisture in the air and the greatest amount of moisture the air could hold at the same temperature. The hotter the grow room is, the more moisture air can hold; the cooler the grow room is, the less moisture the air can hold. When the temperature in a grow room drops, the humidity climbs and moisture condenses. For example, a 800 cubic foot (10' x 10' x 8') grow room will hold a maximum of approximately 14 ounces of water when the temperature is 70°F and relative humidity is 100%. When the temperature is increased to 100°F, the same room will hold approximately 56 ounces of moisture at 100% relative humidity. That's about 4 times as much moisture.

A 10' x 10' x 8' (800 cubic feet) grow room can hold:

4 oz of water at 32°
7 oz of water at 50°
14 oz of water at 70°
18 oz of water at 80°
28 oz of water at 90°
56 oz of water at 100°

The relative humidity factor increases when the temperature drops at night. The more temperature variation, the greater the relative humidity variation. Supplemental heat or extra ventilation is advisable at night if temperatures fluctuate more than 15°F

Testing EC, CF and DS PPM

A nutrient meter's main function is to test the electrical conductivity (EC) in solution. EC is the ability of a solution to conduct an electrical current. Dissolved ionic salts create electrical current in solution and the main constituent of hydroponic solutions is ionic salts.

EC is commonly measured in either a) millisiemens per centimetre (MS/CM) or b) microsiemens per centimetre (US/CM).

1 millisiemen/CM = 1000 millisiemens/CM

PPM testers actually measure in EC but then show a conversion reading in PPM. Unfortunately, the two scales (EC and PPM) are not directly related, because each type of salt gives a different electronic discharge reading. A standard is selected which assumes "so much EC means so much salt" in a nutrient in the solution. The result gives only an idea of the PPM in the nutrient solution.

Conversion scale from PPM/CF

EC MS/CM0.5	Truncheon 0	CF
0.1	70 ppm	1
0.2	140 ppm	2
0.3	210 ppm	3
0.4	280 ppm	4
0.5	350ppm	5
0.6	420 ppm	6
0.7	490 ppm	7
0.8	560 ppm	8
0.9	630 ppm	9
1.0	700 ppm	10
1.1	770 ppm	11
1.2	840 ppm	12
1.3	910 ppm	13
1.4	980 ppm	14
1.5	1050 ppm	15
1.6	1120 ppm	16
1.7	1190 ppm	17
1.8	1260 ppm	18
1.9	1330 ppm	19
2.0	1400 ppm	20
2.1	1470 ppm	21
2.2	1540 ppm	22
2.3	1610 ppm	23
2.4	1680 ppm	24
2.5	1750 ppm	25
2.6	1820 ppm	26
2.7	1890 ppm	27
2.8	1960 ppm	28
2.9	2030 ppm	29
3.0	2100 ppm	30
3.1	2170 ppm	31
3.2	2240 ppm	32
3.6	2520 ppm	36

Hydroponics
Indoor Horticulture

Appendix 3

Glossary

Acid: a bitter or sharp substance, having a pH less than 7 e.g. soil that has a pH of less than 7.5 is considered to be acidic.

Aeration: providing air or oxygen to soil and roots.

Aeroponics: the cultivation of plants by misting the roots suspended in the air.

Air roots: furry, hair-like white roots especially adapted to take up oxygen.

Alchemy: medieval fore-runner of chemistry particularly in the conversion of common elements into gold.

Algae: collective name given to plants that grow extensively in water.

Alkaline: having a pH greater than 7, e.g. soil that has a pH of 7 or greater is considered to be alkaline.

Alkaloids: basic organic compounds containing nitrogen.

Alternating current (AC): an electrical current that alters direction several times a second.

Amino acid: building block of protein.

Ampere (amp): the unit used to gauge the strength of an electrical current.

Annual: a plant that germinates from seed, matures and produces new seeds in a year or growing season e.g. tomatoes are annuals.

Apical: of, at, or being the apex.

Aquaculture: another term for hydroponics.

Atom: the smallest particle of matter possessing the properties of an element.

Auxin: plant growth-regulating substance which, among other things, encourages cell growth by elongation, stimulates cell division thereby encouraging growth, promotes phototropism by growth towards a light source, stimulates fruit development, encourages callouses or basal swelling in wounds, promotes rooting of cuttings, regulates plant height and induces flower formation.

Axil: a location on a plant stem which is the angle between the leaf and the upper part of the stem.

Axillary: arising in the angle of a leaf or a bract.

Bacteria: tiny, unicelled organism.

Ballast: a unit that controls the current of electricity and fires up a HID lamp.

Beneficial insect: a useful insect that preys on harmful plant eating insects.

Biodegradable: the ability to breakdown inorganic and organic substances by natural biological action.

Biosynthesis: the production of a chemical compound by a plant.

Blight: diseases caused by microorganisms where the entire plant is infected and dies.

Blind water or salts: background conductivity of normal water.

Bloom: to produce flowers.

Bloom boost: plant food high in phosphorous that increases yields.

Bolt: an unusual lengthening of plant stems due to elongation of cells.

Bonsai: a very short or dwarfed plant.

Brassica: any member of the family Brassicaceae or Cruciferae, particularly members of the genus Brassica, e.g. cabbage, swede etc.

Bud rot: a wasting disease that affects buds.

Buffer 7: calibration solution for digital pH meters.

Buffering: a chemical substance that stabilises the pH of a solution.

C3, C4: a grouping of higher plants related to the carbon content of the compound produced when carbon dioxide is fixed during photosynthesis.

Calibrate: (1) mark (a gauge or instrument) with a standard scale of readings (2) compare the readings of an instrument with those of a standard.

Calyx: the pod containing the female ovule and 2 prominent pistils, or seed pod.

CAM: crassulacean acid metabolism; a method of photosynthesis found in certain succulent plants that live in hot, dry climates and close their stomata during the day to avoid excessive water loss, and open them at night.

Cambium: a group of actively dividing cells found in the vascular bundles of roots and stems whose function is to produce new plant tissue for lateral growth.

Canopy: the branches and leaves of a woody plant, particularly trees, forming the uppermost light restricting area some distance above the ground.

Carbon: the element which is the basis of organic structure.

Carbon dioxide (CO_2): a colourless, odourless gas, heavier than air which is vital to plant life.

Carbohydrate: a family of organic molecules made up of carbon, hydrogen and oxygen and include sugar, starch and cellulose.

Casparian strip: a waterproof thickening of the radial (side) and end walls of endodermal root cells which is thought to influence the route by which water passes from the cortex into the vascular bundle of the stele.

Caustic: able to burn or destroy living tissue by chemical activity.

Cell: the structural unit of most organisms; cells are bound by a membrane and contains a nucleus.

Cellulose: the main constituent of plant cell walls and the most common organic compound on earth. It fortifies a plant; tough stems contain a lot of cellulose.

Celsius: a scale for measuring temperature where 100° is the boiling point of water and 0° is the freezing point of water.

Centigrade: relating to the Celsius scale of temperature.

CF: conductivity factor of dissolved salts or nutrients expressed as a digital reading.

CFM: cubic feet per minute.

Chelate: the ability for plants to take up metal ions such as iron.

Chemiculture: another term for hydroponics.

Chlorophyll: a group of pigments giving a green colour to most plants, which is found in any part of the plant exposed to light. Chlorophyll pigments are usually contained in the chloroplasts of cells. It has the vital function of absorbing light energy.

Chlorine: a poisonous, green, gaseous chemical element used to purify water.

Chloroplast: containing chlorophyll and found within the cells of plant leaves and stems.

Chlorosis: a yellowing of leaves caused by lack of chlorophyll pigment due to a mineral deficiency, usually iron or magnesium, or by disease (virus) which results in a decrease in photosynthetic rate.

Chlorotic: describing a plant that has the condition of chlorosis.

Climate: the general meteorological conditions prevailing in a given area.

Clone: (1) a rooted cutting of a plant (2) two or more plants with identical genetic make-up produced from one parent by asexual propagation.

CO_2: carbon dioxide formula.

CO_2-enrichment: adding CO_2 to the environment of a grow room to enhance growth.

Colour spectrum: the band of colours produced by separating light into elements with different wavelengths.

Combination roots: a mixture of tap water roots and furry air roots.

Compost: organic matter which is allowed to break down through decomposition. It is high in nutrients and after sufficient decomposition, releases nitrogen.

Concentrate: very refined chemical which needs dilution on application.

Conductivity standard: calibration solution for digital CF meters.

Cork cambium: a specific cambium that contributes towards the production of bark.

Cortex: the layer of plant tissue outside the vascular bundles but inside the epidermis.

Cotyledon: first leaves that appear on a plant after germination.

Cross-pollinate: the transfer of pollen from the anthers of one flower to the stigma of another by the action of wind, insects etc with the subsequent formation of pollen tubes.

Crystal: (1) the appearance resin has when located on leaves (2) many compounds come in soluble crystal form.

Cubic foot: volume measurement in feet; width x length x height = cubic feet.

Cultigens: type of plant that depends on human assistance for its procreation.

Cultivation: the planting, care and harvesting of crops.

Cure: (1) slow drying process i.e. tobacco (2) to make a sick plant healthy.

Cuticle: a thin noncellular layer secreted by the epidermis and prevents water loss.

Cutting: a method of artificial propagation of plants where a small stem, with attached leaves, is removed at a node from a parent plant, and placed in water or a moist growing medium.

Cystine: an amino acid residue formed by oxidation.

Damping off: a disease caused by a fungus e.g. pythium in which young seedlings rot and fall over at ground level in overcrowded, damp conditions; overwatering is the main cause of damping off.

Dead salts: background water reading of dissolved salts in normal water.

Decompose: the breakdown of organic material by rotting or decaying.

Dehydrate: the process by which water or moisture is removed from foliage (or any other substance).

Desiccate: cause to dry up or remove moisture.

Detergent: a substance that when dissolved in water acts as a cleansing agent, pesticide or wetting agent. – NB detergent must be organic to be safe to plants.

Diapause: a period of arrested growth and development in insects which is under the control of the endocrine system. It is an adaptation to avoid adverse conditions.

Dioecious: In plants, it is having male flowers carried by one individual and female flowers carried by another.

Direct current (DC): an electrical current flowing in one direction only.

Disease: an abnormality of an animal or plant caused by a pathogenic organism or the deficiency of a vital nutrient that affects performance of the vital functions.

Dissolve: to make liquid; disperse in a liquid so as to form a solution e.g. crystals dissolve in water.

Dose: (1) known chemical, nutrient or substance (2) a treatment, e.g. amount of insecticide given generally as a solution.

Drainage: removal of excess water; water travels evenly through the growing medium thereby encouraging plant growth. The opposite is where water stagnates in the medium effectively drowning the roots.

Drip Irrigation: a hydroponics technique.

Drip system: an efficient watering system that utilises a hose with small water emitters. Water is dispensed a drop at a time via the emitters.

Ebb and flow: type of hydroponics technique also known as flood and drain and ebb and flow.

Electrode: a conductor through which electricity enters or leaves something.

Electron: a negatively charged subatomic particle found in all atoms and acting as the primary carrier of electricity in solids.

Element: a pure substance consisting of only one type of atom that cannot be destroyed by normally available heat or electrical energy.

Elongate: grow in length, make or become longer.

Embolism: blockage in the vessels such as an air bubble.

Emit: to give off, discharge or send out light.

Endodermis: a one cell layer of tissue found outside the vascular areas and is particularly important in roots where endodermal cells are thickened with casparian strips for control of water transport. Water and salts have to pass through the endodermal cells which act as a kind of valve.

Entropy: the amount of disorder or degree of randomness of a system.

Envelope: outer insulating casing of a lamp.

Enzyme: a protein molecule that catalyses a biochemical reaction by lowering the activation energy required for the reaction to proceed.

Epidermal cells: in stems and leaves, these cells secrete a cuticle.

Epidermis: in plants, this is the thin tissue, usually one cell thick, that surrounds young roots, stems and leaves.

Esoteric: intended for or understood by only a small number of people with a specialised knowledge.

Evaporation: the physical change when a liquid becomes a gas that usually requires heat; this heat is drawn from the immediate environment which produces a cooling effect. In plants, this occurs from the surface of the cells during transpiration.

Evaporating: the action of evaporation.

Exponentially: an increase that becomes more and more rapid.

F_1: first generation of filial offspring of a particular cross.
F_2: second generation, or grandchildren of a particular cross.
F_3: third generation, or great-grandchildren of a particular cross.
F_4: fourth generation, or great great-grandchildren of a particular cross.

Fahrenheit: a scale of temperature in which water freezes at 32° and boils at 212°.

Feed: fertilise, provide nutrients.

Female: having pistils or ovules and seed-producing.

Fertilisation: (1) the fusion of male and female and (2) to apply nutrients.

Fertilise: (1) to apply nutrient to roots and foliage (2) to impregnate female ovule with male pollen.

Fertiliser burn: over fertilisation; leaf tips burn and curl.

Flood and drain: type of hydroponics technique also known as ebb and flow and flood and drain.

Flower: blossom, a mass of calyxes on a stem, top or bud.

Fluorescent lamp: electric lamp made up of a single or double tube(s) and is coated with a fluorescent material. This has low lumen and heat output.

Flushing: clean by passing large quantities of water through.

Foliage: the leaves of the plant.

Foliar feeding: a method of supplying plants with nutrients by spraying an aqueous solution onto the leaves.

Fruit: a plant structure consisting of one or more ripened ovaries, with or without seeds, together with any flower parts which may be associated with ovaries.

Fumigation: disinfect with the fumes or smoke of certain chemicals.

Fungicide: any substance that kills fungi.

Fungus: an organism that lacks chlorophyll. Mildew and mushrooms are examples of fungi.

Fuse: a safety device consisting of a strip of wire that melts and breaks an electrical circuit if the current goes beyond a safe level.

Gas: an air-like fluid substance which expands to fill any space available.

Gene: fundamental physical unit of heredity that transmits information from one cell to another and therefore from one generation to another.

Genetic: of or relating to genes.

Genetic make-up: the genes inherited from parent plants and dictates many important characteristics such as vigour and height.

Genome: the complete complement of genetic material in a cell or carried by an individual.

Genus: a taxon immediately above that of a species and containing usually a group of closely related species.

Germination: the beginning of the growth of a seed, spore or other structure that is dormant.

Growth: the process of increase in size.

Gypsum: a mineral.

H_2O_2: formula for hydrogen peroxide.

Herbicide: any chemical that kills plants.

Hermaphrodite: any plant having both male and female flowers.

Hertz (Hz): a unit of a frequency measuring cycles per second.

HID: High Intensity Discharge.

Honeydew: the sugary waste substance passed out by aphids, scales and mealybugs and secreted onto foliage.

Hood: reflective cover of a HID lamp.

Hormone: a chemical that is produced in the plant and controls growth and development. Root hormones promote the rooting of cuttings.

Horticulture: the science and art of growing plants.

Humidity (relative): the comparative amount of moisture in the air and the largest amount of moisture the air could contain at the same temperature.

Humus: organic material that is derived from the breakdown of plant and animal material occurring in the surface layers of the soil. Humus is black in colour, very fertile and a soil improver.

Hybrid: an offspring of a cross between two genetically different individuals.

Hybridisation: artificial creation or breeding of two different species or varieties to produce hybrids.

Hydrogen: light, colourless, odourless gas; it blends with oxygen to make water.

Hydroponics: the science of growing plants without soil, in which the roots are suspended in aerated water containing known quantities of chemicals that can be adjusted to suit changing conditions.

Hydroponicists: people who grow using hydroponics.

Hydroponicum: hydroponic installation.

Hygrometer: instrument for measuring relative humidity.

Imperial: (of weights and measures) conforming to a non-metric system formerly used in the UK.

Induce: cause or influence by stimulation i.e. flowering can be induced via a 12 hour photoperiod.

Inert: chemically non-reactive, for example inert growing media. Controlling the nutrient solution is made easy because the growing medium is inert.

Insecticide: any substance that kills insects.

Intensity: the extent of light energy per unit; intensity decreases the further away from the source.

Internode: the portion of a stem found between 2 successive nodes (see Nodes).

Ion: an atom that carries a charge due to loss or gain of electrons.

Jiffy-7 pellet: compressed peat moss, wrapped in plastic. When moistened, it expands into a small pot that is used to start seeds or cuttings.

Kelvin: an absolute scale temperature where a range of 1 kelvin (K) is equal to a range of 1°C.

Leach: dissolve or remove nutrients or other substances from growing mediums by the percolation of water.

Leader: see Meristem.

Leaf: the principle photosynthetic organ of vascular plants which typically consists of a flattened lamina joined to the stem by a stalk or petiole at which junction an axillary bud can be found.

Leaf blade: that part of the leaf where photosynthesis takes place.

Leaf curl: leaf malformation due to over watering, over fertilisation, lack of magnesium, or insect or fungus damage.

Leaflet: small immature leaf.

Leaf whorls: a circular set (2 or more) of leaves arising at the same level on the plant.

Leggy: having long and straggly stems with little foliage usually caused by lack of light.

Life cycle: the serial progression of stages a plant must pass through in its natural lifetime; for example in an annual plant, it passes from seed, seedling, vegetative and floral.

Lime: used to raise and stabilise pH.

Litmus paper: paper used for testing pH.

Lumen: measurement of light output; one lumen is equal to the amount of light emitted by one candle that falls on one square foot of surface located one foot away from one candle.

Lux: measurement of lumens or light.

Macronutrient: any element required in large quantities for growth such as the primary nutrients N-P-K.

Mandibles: one of a pair of the mouthparts of an insect used for crushing food.

Manicure: trimming of foliage and stems from plants.

Mean: the average of.

Medium: a substance in which plants can be grown. A medium can be liquid or solid. It can contain all the necessary nutrients and trace elements for normal growth, or these can be supplemented.

Metabolism: the sum total of chemical processes occurring in cells by which energy is stored in molecules or released from molecules, life being maintained by a balance between the rates of processes.

Meristem: a region of a plant in which active cell division occurs. They are located at the tip of plant growth.

Micronutrients: any trace elements or compounds which include, S, Fe, Mn, B, Mo, Zn and Cu.

Micro-organisms: any microscopic organism such as bacteria or fungus.

Mineral salt: any inorganic homogeneous solid such as sodium, potassium, phosphorous, chlorine.

Mist: to administer water with the aid of a spraying mechanism.

Molecule: the smallest chemical unit of matter that has the characteristics of the substance of which it forms a part.

Mother plant: female plant grown for cuttings or cloning stock and kept in a perpetual vegetative state.

Mulch: a protective covering that can be compost, old leaves, straw etc; however it can cause fungus outbreak as the medium is kept too wet.

Mycelium: the total mass of hyphae of a fungus that constitutes the vegetative body (as opposed to a fruiting body).

Mylar: a reflective sheeting material.

Necrosis: localised death of a plant part.

NFT: abbreviation for nutrient film technique.

Nitrogen (N): essential to plant growth and one of the three primary macronutrients.

Nodes: the part of the plant stem where leaves are attached or may develop from buds.

N-P-K: Nitrogen, Phosphorous, Potassium.

Nucleic acid: a molecule comprising a sequence of nucleotides forming a polynucleotide chain. Nucleic acids act as the genetic material of cells and occur as either DNA or RNA.

Nucleus: an organelle of eukaryotic cells that is bounded by a nuclear membrane and contains the chromosomes whose genes control the structure of proteins within the cell.

Nutriculture: another term for hydroponics.

Nutrient: substance that plants take in and assimilate for growth and maintenance; needs essential elements N-P-K and secondary and trace elements which are fundamental to plant life.

Nutrient lock: when a conflicting concentration of nutrients mix together.

O_3: formula for ozone.

Organic: made of, or derived from or related to living organisms. In agriculture, organic means 'natural'. In chemistry, organic means 'a molecule or substance that contains carbon'.

Osmosis: the movement of a solvent through a differentially permeable membrane from a solution with high water concentration and low solute concentration to one with a low water concentration and high solute concentration.

Ovipositor: an organ of female insects, usually present at the tip of the abdomen through which the egg is laid. It is sometimes developed to enable the piercing of tissues, particularly where eggs are laid inside other insects, animals or plants.

Ovule: a structure found in higher plants that contains an egg cell and develops into a seed after fertilisation.

Oxidation: the addition of oxygen to a substance to increase the proportion of oxygen in its molecule. Oxidation can be achieved without oxygen by the removal of hydrogen.

Oxygen: tasteless, colourless gas and forming about 21% of earth's atmosphere, that is capable of combining with all other elements except the inert gases. It is necessary to maintain plant and animal life.

Ozone: a strong-smelling, poisonous form of oxygen, formed in electrical discharges or by ultraviolet light.

Ozone generator: a machine that generates ozone.

Palisade: or also called mesophyll. The internal tissue of a plant leaf except the vascular bundles. All mesophyll cells contain chloroplasts for photosynthesis which lie close to the edge of the cell in order to gain maximum light and gas supply. Mesophyll tissue contains numerous intercellular spaces which communicate with the atmosphere outside the leaf via the stomata.

Parasite: organism that lives in association with and at the expense of another organism, the host, from which it obtains organic nutrition.

Parenchyma cells: these are thin walled, general purpose plant cells that usually have a packing function.

Passive: hydroponic system that moves the nutrient solution passively through absorption or capillary action.

Pathogen: any organism that causes disease such as a virus, bacterium or fungus.

Peat: an accumulation of dead plant material formed in wet conditions in bogs or fens in the absence of oxygen so that decomposition is incomplete. It is usually acidic.

Percy throwers: slang for personal growers i.e. someone that grows solely for oneself.

Pericycle: the layer of plant cells between the endodermis and the phloem consisting mainly of parenchyma which becomes meristematic to form lateral roots.

Perlite: (1) sand or volcanic glass expanded by heat. It holds water and nutrients on its many irregular surfaces (2) mineral soil improver.

Pest: any organism that causes nuisance to man either economically or medically. The classification of pests is forever changing and is very dependent on economic circumstances.

Pesticide: an agent that causes the death of a pest. The general definition is usually restricted to pesticidal properties such as herbicides, insecticides and fungicides.

Petiole: the stalk of a leaf containing vascular tissue which connects with the vascular bundles of the stem. The base of the petiole where it joins the stem, may have small leaflike structures called stipules and axillary buds.

pH: a measure of the hydrogen ion concentration in an aqueous solution on a scale from 1-14, 1.0 being very acidic, 14 is highly alkaline and 7.0 being neutral. Plants grow best in a range of 5.5 to 6.8.

pH tester: electronic or chemical indicators used to determine the pH level in the growing medium or water.

Phloem: a transport tissue characterised by the presence of sieve tubes, companion cells and phloem parenchyma cells found in the vascular bundles of higher plants. Phloem functions in the transport of dissolved organic substances e.g. sucrose by translocation.

Phosphor coating: internal bulb coating that diffuses light and is responsible for various colour outputs.

Phosphorous (K): one of the three macronutrients that promotes root and flower growth.

Photometrics: the study of light, particularly colour.

Photon: a quantum of radiant energy with a wavelength in the visible range of the electromagnetic spectrum.

Photoperiod: the length of daylight as compared with the length of darkness in each 24 hour cycle.

Photosynthesis: the process by which plants convert carbon dioxide and water into organic chemicals using the energy of light with the release of oxygen.

Phototropism: a bending growth movement of parts of a plant in response to light stimulus. The movement produced by unequal growth is due to differences in auxin concentration.

Pigment: (1) the substance used for colouring anything and that absorbs light reflecting the same colour as pigment (2) a natural substance that gives plant or animal its colour.

Pistil or carpel: the flask shaped female reproductive unit of a flower made up of ovary, style and stigma.

Pith: core of a dicotyledon stem and contains parenchyma cells which have a storage function.

PK13/14: a specific bloom boost nutrient.

Pollen: a small structure of higher plants that is yellow and dust-like which contains male gamete nuclei and is surrounded by a double wall. Pollen is transported from the male stamen to the female stigma by pollination.

Power surge: interruption or change in intensity of electricity.

PPM: parts per million.

Predator: any beneficial insect, parasite or animal that lives by hunting down and eating other harmful insects.

Primary nutrients: N-P-K.

Propagate: (1) sexual: produce a seed by breeding different male and female flowers; (2) asexual: to produce a plant by artificial or natural means.

Propagator: a container designed to generate maximum humidity for seeds or cuttings.

Propane: a flammable gas present in natural gas and used as a bottled fuel.

Protoplasm: the living contents of a cell, i.e. the cytoplasm and nucleus.

Prune: trim and modify the shape and growth pattern of a plant.

Pukka: colonial (Hindi) expression for excellent.

Pupate: an apparently inactive phase between larva and adult insect. Although movement and feeding are absent, extensive developments take place in the formation of adult structures.

Pyrethrum: a natural insecticide prepared from the dried flowers of the chrysanthemum plant.

Pythium: fungal disease.

Radicle: the basal part of the embryo in a seed, developing into the primary root of the seedling.

RCD: Residual Current Device – measures current and a circuit breaker.

Reflector: the housing for a HID lamp, also known as a hood.

Regeneration: the replacement of tissues or repair of tissues or organs lost through damage.

Rejuvenate: restore youth; a mature plant having concluded its lifecycle (flowering) may be reinvigorated by an 18 hour photoperiod in order to produce new vegetative growth.

Resin glands: tiny pores or ducts that produce resin or oils.

Respiring: carry out respiration; a process by which gaseous exchange (oxygen and carbon dioxide) takes place between an organism and the surrounding medium.

Re-translocating: the ability to locate, translocate then re-translocate, relating to the movement of stored nutrients in a plant.

Reverse osmosis (RO): water purifying technique.

Rockwool: an inert, sterile growing medium made from fibreglass.

Root: (1) the part of the plant which usually grows below ground. It provides anchorage for the aerial parts, absorbs water and mineral salts from the growing medium, conducts water and nutrients to other parts of the plant and often stores food materials over winter. (2) to root a cutting or clone.

Rootball: entire below ground vegetative system of the plant.

Root cap: a structure found at the apex of all roots (except those of many water plants) produced by apical meristem. The cap consists of a thimble shaped collection of parenchyma cells which have a protective function. As the root pushes its way through the soil, the outer or older cells of the root cap are sloughed off and replaced by new cells from the meristem.

Root hair: a hair-like outgrowth from the epidermis of roots. Root hairs occur in large numbers in a zone behind the growing tip, are short lived and greatly increase the absorbing area of the root.

Root hormone: root-inducing hormone.

Root zone: entire root system of plant, also known as rootball.

Rooting compound: chemical hormone developed to aid the rooting process of a cutting.

Salt: crystalline compound that results from pH or toxic build-up of nutrient. Salt will damage plants and prevent them from absorbing nutrients. Minerals salts are nutrients that are utilised in hydroponic formulas.

Secondary nutrients: calcium (Ca) and magnesium (Mg).

Seed: the structure formed in the fertilised ovule of a plant consisting of an embryo surrounded by a store for nourishment during germination.

Shaman: a person believed to be able to contact good and evil spirits.

Shoot: that part of a vascular plant which is above ground consisting of stem and leaves.

Soap: (1) cleaning agent (2) wetting agent (3) insecticide. It should be biodegradable.

Solenoid: an electronically controlled on/off valve.

Soilless: a growing medium made up of mineral particles such as vermiculite, sand, pumice, but not soil.

Soluble: able to be dissolved in another substance e.g. water.

Solution: (1) a homogenous mixture in which a substance (solid, liquid or gas) is dissolved in another substance usually of solids in liquids. (2) answer to a problem.

Spore: a reproductive body consisting of one or several cells formed by cell division in the parent organism which when detached and dispersed and conditions are right, will germinate into a new individual. Spores occur particularly in fungi and bacteria and protozoa. Some have thick resistant walls and can overcome adverse conditions such as drought. They are usually produced in large numbers and occur as a result of sexual or asexual reproduction.

Stagnant: motionless air or water; water must be in motion in order to promote healthy plant growth.

Stamen: the organ in flowers consisting of a stalk bearing an anther in which pollen is produced.

Starch: a complex carbohydrate; starch is the principle storage compound of plants.

Stem: the part of the shoot of vascular plants from which are produced leaves at regular intervals (nodes) and reproductive structures. Stems are usually circular in cross section but some are square while others are ribbed.

Sterile: (1) free from living micro organisms (2) not capable of producing offspring.

Sterilise: the removal of dirt, germs and bacteria by destroying all forms of microbial life.

Stoma (singular), stomata (plural): an opening in the epidermis of leaves and sometimes the stems that allows gaseous exchange.

Strain: (1) a group of organisms within a species or variety distinguished by one or more minor characteristics (2) to abuse a plant by withholding nutrients, water etc.

Stretch: increase in length.

Stress: a physical or chemical factor that puts pressure on plants.

Stripping: remove leaves from plant.

Suberin: a complex of fatty substances present in the wall of cork tissue that waterproofs it and makes it resistant to decay.

Substrate: a medium on which an organism can grow.

Sucker system: (1) an axial shoot arising from an adventitious bud of a root and appearing separate from the mother plant (2) a new shoot on an old stem (3) a modified root in a parasite enabling it to absorb nutritive materials from the host.

Sugar: a simple form of carbohydrate and food product of a plant.

Symbiosis: a relationship between dissimilar organisms in which both partners benefit.

Synthesis: production of a substance such as chlorophyll where solar light energy is incorporated into complex compounds such as glucose.

Tannins: a complex organic compound occurring widely in plant sap, particularly in bark, leaves and unripe fruits.

Taproot: a primary root structure in which one main root forms the major part of the underground system; lateral roots will branch off the taproot.

TDS: abbreviation for total dissolved salts, type of measurement value like PPM and CF.

Terminal bud: carried at the growing end of the stem.

Thermostat: a device for regulating temperature to control a heater or a fan.

Thiamine: or vitamin B1: a water soluble organic compound found in cereals and yeast whose main function is to act as a coenzyme in sugar breakdown by forming part of the NAD molecule.

Timer: an electrical mechanism for adjusting photoperiods, for example a fan. A timer is essential in grow rooms.

Tip: point or end of leaf.

Tip burn: when the very end tip of the leaf turns brown.

Tissue: any group of cells of similar structure in animals or plants that performs a specific function.

TMV: Tobacco Mosaic Virus; a rod-shaped viral particle consisting of an RNA helix surrounded by a protein coat. The virus was the first to be purified and crystallised. It causes yellowing and mottling in the leaves of tobacco and other crops (e.g. tomato) which adversely affects the quality and quantity of crop produced.

Trace element: any element that is necessary for the proper working of biological systems. Absence can cause disease and death.

Transformer: a device in the ballast that alters electric power from one voltage to another.

Transpire: the loss of water vapour and by-products from inside the leaf via the stomata. It exerts considerable upward pressure in the stem and is thought to explain how water goes up from the roots and into the leaves.

Ultraviolet: type of electromagnetic radiation beyond the visible spectrum of violet light.

Unicellular: or acellular. An organism having a body which is not composed of cells and is quite complex in structure.

Variety: a group of organisms that differs in some way from other groups of the same species.

Vacuole: a membrane-bound compartment within the cytoplasm of a cell, containing cell sap, for example water, air, food.

Vascular: (of vessels) conducting fluid, e.g. water in plants. Vascular bundle is a structure of vascular plants that runs up through the roots, into the stems and out into the leaves and whose function is transport within the plant.

Vascular cambium: a group of actively dividing cells found in the vascular bundle between the phloem and xylem.

Vegetative: growth of or relating to the non-sexual organs of a plant such as root, stem and leaves and is the growth stage of plants when they are producing new growth and chlorophyll very rapidly and abundantly.

Ventilation: circulation of fresh air, essential for healthy plants. Fans create ventilation.

Ventura: a scientific principle relating to the effects of air and water when under water.

Vermiculite: mica processed and expanded by heat. Vermiculite is a good soil improver and growing medium for rooting cuttings.

Water roots: type of root specially designed for water uptake, also known as taproots.

Weeds: any undesirable plants and growing in the wrong place at the wrong time.

Wetting agent: compound that diminishes the droplet size and reduces the surface tension of the water thereby making it wetter. Washing up liquid is a good wetting agent, if it is biodegradable.

Wick: used in passive hydroponic systems by utilising a wick that is suspended in the nutrient solution so that the nutrients travel up the wick and are absorbed by the medium and roots.

Yo-yo: a device for supporting heavy yielding plants.

Xylem: a woody plant tissue that is vascular in function, enabling the transport of water with dissolved minerals around the plant, usually in an upward direction.

Hydroponics
Indoor Horticulture

Index

Hydroponics
Indoor Horticulture

Hydroponics
Indoor Horticulture

Hydroponics
Indoor Horticulture

Hydroponics
Indoor Horticulture

Hydroponics
Indoor Horticulture

Hydroponics
Indoor Horticulture

Hydroponics
Indoor Horticulture

Hydroponics
Indoor Horticulture

Hydroponics
Indoor Horticulture

Hydroponics
Indoor Horticulture